D0915896

The Restless Plant

THE RESTLESS PLANT

DOV KOLLER

Edited by Elizabeth Van Volkenburgh

Harvard University Press
Cambridge, Massachusetts, and London, England   2011

Library of Congress Cataloging-in-Publication Data

Koller, Dov.
The restless plant / Dov Koller ; edited by Elizabeth Van Volkenburgh.
p.   cm.
Includes index.
ISBN 978-0-674-04863-8 (alk. paper)
1. Plants—Irritability and movements.
2. Tropisms.   I. Van Volkenburgh, Elizabeth.   II. Title.
QK771.K58 2011
575.9'7—dc22   2010013249

In memory of Dov Koller (1925–2007)—
scientist and teacher; husband, father, and grandfather

Contents

Foreword

This book is the product of two of my father's passions. The first was his passion, as a researcher, for understanding the miracle of plant movement, a topic that was the focus of his research for over 25 years. The second was his passion for teaching—for reaching out to people and conveying the beauty found in science and nature. My father was struck by the fact that, although plants are all around us, relatively few people realize how complex and rich the plant world is. Although the remarkable phenomenon of plant motion was first explored by Darwin in 1880, this concept and all the rich understanding that has emerged since Darwin's revelation are nevertheless unfamiliar to most people. This book is my father's attempt to make these ideas accessible to more people, and thereby reveal to them another facet of the beauty of the world.

This book consumed my father for two years. He worked on it day and night, even in the hospital where he spent a large part of the last months of his life. Days before his death, he produced a complete draft of the text, but one that was lacking figures, references, and an index. On behalf of my family, we would like to thank the people who helped, both before and after my father's death, to bring this book to completion. Professor Winslow Briggs and Professor Jaime Kiegel had many useful discussions with my father in the process of writing the book. Professor Roger Hangarter and the anonymous reviewer made valuable comments on the first draft of this manuscript. Dov Bock produced the beautiful illustrations that help visualize the concepts in the book. Professor Alfred Mayer stepped forward after my father's death and generously provided tremendous help in detailed comments and suggestions regarding the figures and references.

We are very grateful to Ann Downer-Hazell and Michael Fisher from Harvard University Press for their constant support and their willingness to take a risk on a book without a living author. But most of all we are deeply grateful to Professor Elizabeth Van Volkenburgh, who agreed to take on a partially completed manuscript and turn it into a book that embodies my father's vision. We believe that the final product would have made him very proud.

Daphne Koller, Stanford University
on behalf of the family

Author's Preface

The Restless Plant is my comprehensive, up-to-date account of the incessant movements in all actively growing plants. Developments and advances in this area of plant sciences since the 1880 publication of Darwin's *The Power of Movement in Plants* have been tremendous. This is not a textbook, nor is it a scientific review. It is intended for the general, educated lay public of all scientific and nonscientific disciplines as well as for that part of the general public who might be interested in the endless variety and fascinating phenomena of plant movements, and in what is known about what makes them happen.

I have attempted to make the book as user friendly as possible by using the plainest language, but some of the accounts are inevitably somewhat scientific and technical. I have also avoided all references to published literature and experimental data in tables and graphs. Quoting or citing them in this book would make it unwieldy and unpalatable. At the same time, because I wanted to make the book as useful as possible, I have provided sections to explain the fundamental processes involved in the operation of plant movements and their motor organs, and the basics of the interactions between matter and the electromagnetic energy of light. I have provided a comprehensive glossary to define and explain, successfully, I hope, the scientific terms as they appear in the text.

Where I saw fit, I offered my own interpretation of the mechanism and adaptive significance of plant movements. I realize that I may have overemphasized areas that I have found more significant than others. I also realize that some of my interpretations may not agree with accepted dogma. Readers wishing to reach their own interpretations and conclusions by pursuing the actual research are referred to review articles, encyclopedias, the Internet, and relevant textbooks on the plant sciences as starting points.

Editor's Preface

Plants are curious organisms—clearly they are alive, but their apparent lack of any central organization (brain, heart, or nervous system) makes it harder for us to understand how they work. They are slow: usually they look today just like they did yesterday, and we have to wait, sometimes a long time, to see changes in them. Their metabolism, development, and sensory and response behaviors are orders of magnitude slower than those of most animals. Their bodies and development are not centralized but rather are diffuse; they grow in modular units, and often detached parts can reestablish as individuals (natural cloning) or be replaced. Plants don't talk to us, so we are left to decipher what they are up to. This is the nature of inquiry, experimentation, and discovery.

As biological organisms, a plant's life imperative is the same as that of all others: to optimize reproductive fitness, that is, to maximize the number of viable offspring. We are familiar with angiosperms (flowering plants) and gymnosperms (cone-bearing plants), with ferns, mosses, algae, and other photosynthetic organisms. We garden with them, we eat them, and we concern ourselves with agricultural policies and the health of our global ecology. Astonishingly, however, discoveries that were made about plants as long ago as several centuries have barely percolated into general awareness and education. Plants reproduce sexually; plants contain DNA and inherit traits genetically, as observed by Mendel and others since. Yet only decades ago did mention of plant "sex" become common in the classroom, and it is not unusual to see produce in the market labeled "NO DNA." Plants sense their environment, and they coordinate sensory information into responses that optimize their growth and reproduction—phototropism was studied by the Darwins in the 1880s, and they identified the site of a seedling's "vision" at the tip of its coleoptile or hypocotyl. Plants like sunflowers reorient every morning to face the sun by "memory." We can definitely say plants sense and respond; can we also say they behave? Plants move, and they sense their position and adjust it underground as well as above. Their anatomy is well understood; there are no muscles—but there is actin and myosin in plant cells; there are no bones, but there is wood; and there are no nerves, but there are long-distance electrical and other signals running through the vasculature.

Dov Koller brings to life these often underestimated, absolutely necessary life-supporting activities of plants. With scientifically accurate attention to detail, and occasional mathematical analysis, he outlines the forces plants create to drive the motors that move plant organs. He notes that plant cells differ from animal cells by two main features creating their hydrostatic structure. (1) A huge central vacuole spreads the cytoplasm into a thin sheet around the cell's periphery (increasing the surface area without wasting protein); the vacuole is critical for establishing hydrostatic pressure to drive many plant motors. (2) Rigid cellulosic cell walls not only provide resistance against vacuoles to create hydrostatic structure, but in wood are the basis for compressive structure (providing the mechanical means to create catapults and other ingenious structures for movement and dispersal). From these basic points, he describes simple, well-known movements, such as growth-mediated curvature of exploratory root tips, phototroping seedlings, and plants growing out from under shade. He gives examples of single-celled motors that are at once so simple yet of utmost importance to plant life—the guard cells bounding the breathing pores, the stomates, and the pollen tube, which navigates through the flower's style to the ovary, carrying sperm nuclei to the egg. But more surprising are his descriptions of the less well-understood movements—those underground and in plant canopies. It may be a surprise to learn that underground plant parts, corms, stolons, roots, and stems sense their position in order to situate themselves at their "physiological depth," sufficiently deep to be safe from hazards at the soil–air interface, but not so deep that they lose access to air and organisms growing at the top of the soil column. Most of us have observed leaves orienting toward the source of light, but how does that work? The location of photoreceptors may be remote from the responding motor that moves the organ—how are these communicating? And canopies—looking up through a Japanese maple canopy is a stunning experience when one realizes that the leaves rarely shade each other. How do they sense and then position themselves to make this mosaic display? This book makes a fundamental point that all students of plant biology can grasp and become fascinated by: many behaviors of plants are created by subtle shifts in the locations, rates, and duration of growth physiology. This is an underlying basis for taxonomic distinctions among plants, their morphology, function, and role among others in biological communities.

In the early 1980s, Dov Koller visited the University of Washington's Botany Department, where he gave an informal talk on leaf movements. At the time, he was on sabbatical leave in Dr. Winslow Briggs's laboratory at the Carnegie Institute in Palo Alto, California, and I was a graduate student at the University of Washington. We gathered in a classroom, sitting at student desks, welcoming a break from the lab bench. Dov was engaging and relaxed, and he made us immediately at home with his topic: how sun-tracking leaves perceive and respond to light. Dov spoke simply, directly. He drew figures on the board and posed questions. He told us what he knew, but more importantly what he didn't know. He challenged our understanding of physics and mechanics, both inside and outside of the plant, by describing experimental methods most of us had never imagined. He was ingenious, practical, and creative. That afternoon he drew a cartoon in my mind by asking, how does a cell in the leaf blade communicate with the pulvinus at its base? The question is simple, but we still don't know the answer. Dov's curiosity, clarity, clever experiments, and appreciation for the beauty of biological sensory behavior made an impact—he taught by inquiry, and his questions were profound.

In this book, he has collected results from his own and many others' experimental investigations into the nature of movement in plants. More than a compendium of information, the book will, we hope, open minds, provoke questions, and propel many to continue pursuing the mysteries that captivated Dov Koller's imagination.

It has been a pleasure for me to immerse myself in this material and to work with Dov Koller's daughter, Daphne Koller, to bring the book to fruition. I would like to thank Daphne for entrusting this task to me, her mother and family for sending materials from Dov Koller's collections of photos, Dov Koller's friend and colleague Alfred Mayer for suggestions for illustrations and references, and Ann Downer-Hazell for persisting. Thanks also go to the PhytoGrads at the University of Washington biology department for their review of the original manuscript and recommendations, and to Dr. Robert E. Cleland for his insight and critical discussion of mechanisms of movement and growth in plants.

The Restless Plant

CHAPTER ONE

A Guided Tour of Plant Movements

NIMALS CLASSIFIED as amphibians are able to exist in water and on land, and they spend their lives shuttling between these two environments. In this respect, all but very few plants are obligate amphibians, with part of their body permanently in the aerial environment and the remaining part within the soil. This structural differentiation in plants is based on function. Nutrients essential for plant growth and development are available in the soil solution and in the air, and they are available in exceedingly low concentrations. Plants require entirely different harvesting techniques and equipment to feed from these physically distinct sources. Roots ramify through the soil, spreading their tips into new territories for gathering nutrients and water. Root hairs and associations with mycorrhizal fungi greatly increase absorptive surface area. Stems and branches grow up into the atmosphere, spreading leaf surfaces out to exchange gases and serve as antennae to catch light for photosynthesis. Both roots and shoots use environmental cues, gravity, and light to guide their growth and movements toward optimizing reproduction, their ultimate goal (Fig. 1).

Terrestrial plants are stationary, incapable of changing their location, yet portions of their body may exhibit clearly visible movements. These movements, whether they are bending toward light, tracking the sun, opening a flower, or pulling a rhizome through the soil, generally take place over a time scale that is orders of magnitude slower than our eyes are accustomed to. For this reason, plants seem to be still while in fact they are always moving, seeking, and changing position. Time-lapse photography captures these movements and, by speeding them up, displays often quite surprising

behaviors (see, for example, Plants in Motion at http://plantsinmotion.bio
.indiana.edu and ChloroFilms at www.chlorofilms.org).

The universally accepted, physiological definition of movement in plants
is restricted to changes in the spatial orientation or the conformation of
an organ or its part(s). Most often recognized are rapid leaf movements
of sensitive plants such as *Mimosa pudica* (Fabaceae) or the carnivorous
Venus' flytrap (*Dionaea muscipula,* Droseraceae) (Fig. 2). A broader understanding of plant function, however, would include displacement of
elongating, expanding, growing organs. Entire leaves, leaflets, root apices,
apical buds, and their subtending cluster of developing leaves, flowers, or

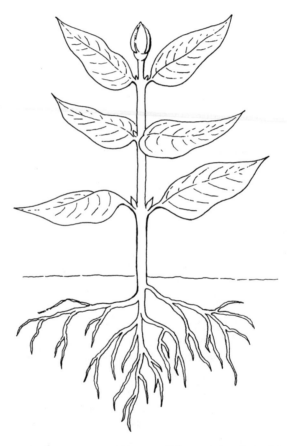

Fig. 1. Plants spread out, roots penetrate soil for water and nutrients, and
shoots and leaves reach into the atmosphere to gather light and carbon dioxide
for photosynthesis. Rooted in space, each organ senses its environment and
signals information internally, adjusting position, function, and other behaviors
accordingly.

inflorescences, fruits, or entire shoots, may be capable of moving. Individual cells may move, such as guard cells regulating stomatal pore size and pollen tubes penetrating the style of a flower. All of these movements are driven by motors powered by internal pressures or tensions exerted against cell walls adapted to the purpose (see Chapter 2, Motors). These cause anisotropic changes in the dimensions of individual cells or cells composing entire tissues and create movements exhibiting a number of distinctive characteristics.

Sensitive Plants

A number of plants with compound leaves (*Mimosa pudica, Cassia fasciculata*, and *M. spegazzini* Leguminosea; *Biophytum sensitivum* Oxalidaceae) are classified as sensitive because they exhibit rapid down-folding of their leaflets when the entire plant or a single leaf is mechanically perturbed, by touch or by shaking. In nature, this behavior may protect the leaves from being eaten by insects and perhaps also from damage during rain or windstorms. In the double-pinnate leaf of *Mimosa*, such movements are driven by motor tissues in pulvinules of the individual leaflets (pinnules), in pulvini at the juncture between the rachilla (pinnae) and rachis, and even in the pulvinus at the base of the petiole (the juncture with the stem). The response takes place as an almost immediate loss of water and turgor in the lower (abaxial) sector of motor tissues of the pulvinus; down-folding after shaking is completed within 1 to 2 seconds. Loss of turgor results from a sharp increase in membrane permeability (plasmalemma and tonoplast), leading to loss of water and solutes from the protoplast into the apoplast and intercellular spaces. Recovery to the original configuration requires uptake of osmotically active solutes (ions) into the vacuole, followed by water; recovery therefore takes place more slowly, but it is nevertheless completed within 15 minutes.

Recovery of the pinnules and pinnae, but not of the base of the petiole, takes place only in light. Light is required to activate plasma membrane H^+-ATPase proton pumps to enable the influx of K^+ accompanied by Cl^- into the cell. Sensors of mechanical perturbation are located in each individual pulvinus, and they are probably also activated by its structural deformation of intracellular organelles. Excitation of the individual sensors appears contagious; it spreads to untouched pulvinules close by and even at some distance. Transmission of the stimulus is probably electric. Shaking the branch or whole plant causes its leaflets to produce torque and deformation at their pulvinule. The responses are classified as seismonastic. Movements associated with traps of certain insectivorous plants (Droseraceae) are driven by structural deformation caused by mechanical perturbation of specialized sensory organs such as bristles (in *Dionaea*) or tentacles (in *Drosera*).

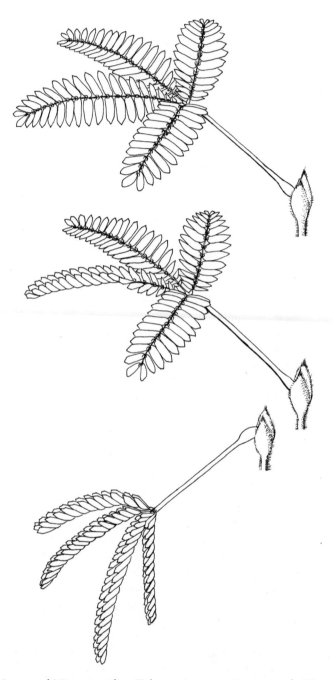

Fig. 2. Leaves of *Mimosa pudica* (Fabaceae) are sensitive to touch. Disturbed leaflets fold up and signal neighboring leaflets, which follow suit. Ultimately, the whole leaf folds downward.

Initiation of Movements

Movements of plants are stimulated both by internal signals and by signals from the environment. Nastic movements take place in genetically defined spatial directions. Such movements are, and their direction is, species specific and predetermined endogenously and exclusively by the fixed location and structural disposition of the driving motor or its parts. Tropic movements take place in directions that are determined by some vectorial environmental signal, such as gravity or light.

Internal Signals

Nastic movements are considered autonomous because they are controlled by signals produced by the plant (or cell) itself. Autonomous signals may be generated by fundamental developmental changes such as germination of seeds, transition from vegetative development to flowering, or onset of dormancy. They may also be generated in the course of normal growth processes. For instance, movement of root hairs into the soil is an integral part of the development of the young root; initiation of leaf primordia and the movements that they make to become part of a bud occur with the development of the young shoot.

Underlying many movements in plants is the universal biological oscillator ("clock"). This mysterious physiological mechanism operates on a circadian (around a day's length) cycle, the timing of which is genetically and autonomously determined. Its function is easily monitored in plants grown under constant environmental conditions (light, temperature, and humidity) where they exhibit autonomous movements with a remarkably constant rhythm. A plant's free-running rhythm can be up to 30 hours and is usually longer than 24 hours. Sleep movements of leaves, opening and closing of stomata, and nighttime opening of flowers are examples of such circadian movements.

A variety of nastic movements are characteristic of dorsiventral organs such as leaves, petals, and sepals. These organs exhibit structural and/or physiological differences between their upper side, designated as ventral or adaxial, and the lower side, designated as dorsal or abaxial. As they expand, such developing organs may exhibit equal expansion on both sides, remaining flat. Alternatively, expansion of one side may be more rapid than the other, resulting in organs that are curved downward (epinastic) or upward (hyponastic). Such bending movements of young leaves are growth mediated, part of their normal development and entirely under endogenous control. The growth is irreversible, but the curvature is reversible. It is thought that epinastic and hyponastic movements result from differential distribution of the growth hormone auxin between the upper and lower side of the organ, although roles for other hormones including ethylene and abscisic acid also have been described. Maintenance of "flatness" in an organ such as a leaf remains a mystery, however, and poses excellent research questions involving not only how regulation of growth occurs physiologically, but also what biomechanical properties govern the developing shape, and what the nature of signaling among developing tissues is such that they create a flat structure.

Buds exhibit spectacular epinastic and hyponastic movements, an inherent feature of their development. All foliage leaves (green or not) are formed from primordia in vegetative buds, and all the reproductive organs of the plant are formed in flower buds. The meristem of the vegetative shoot apex is shaped like a dome, and it produces leaf primordia in an acropetal (base to top) spiral sequence around its base in the form of upright crescents. At first, the leaf primordium grows by producing more cells, and it exhibits hyponasty by more rapid growth of its lower side, which becomes convex, while its upper side becomes concave. As a result, the young, developing leaf cups to form a tight shield around the younger primordia and the apex. Elongation of the internodes between the leaf primordia is suppressed entirely so that hyponasty of the developing leaves leads to the formation of a compact bud. All this is reversed as the lowermost young leaves change over from hyponastic to epinastic growth. This occurs as cells begin to increase markedly in volume. More rapid expansion of the upper part of the leaf changes the configuration of the leaf from convex/concave to the typical flat (coplanar) leaf lamina (blade). At the same time, the dormant internodes start elongating vigorously, separating the just-flattened leaf laminae and transforming them into an integral part of the shoot. Therefore, in the normal course of leaf development, the sequential hyponasty and epinasty is expressed autonomously. The determining factor is the competence of the shoot to promote opening in these buds, and this is likely to be under seasonal control. Opening of the bud enables its leaves to expand and fulfill their function in harvesting light energy.

Flowers of angiosperms are also produced first as buds containing the apical meristems of the shoot. When the vegetative plant receives its signals to undergo transition to a sexually competent, reproductive state, the apical bud stops making leaves and starts making flower parts: sepals forming the calyx; petals forming the corolla; and fertile sporophylls, stamens (male sporophylls), and carpels (female sporophylls) that make the ovary. Ultimately, the function of flowers is to facilitate pollination.

To do that, the stamens and ovary have to be exposed. Their exposure takes place by unfolding of the calyx and corolla, employing epinasty, as in vegetative buds.

In some species, the corolla unfolds once, before senescing and dying. Species of *Oenothera* (Onagraceae) and morning glories (Convolvulaceae) produce such ephemeral flowers that they unfold only once, in early evening, then the corolla withers and dies. In other species, such as the night-flowering cactus *Cereus grandiflorus* (Cactaceae), the flowers unfold and refold repeatedly, employing epinasty as well as hyponasty. In both cases, the movements are not controlled by the diurnal light/dark cycle but rather are under circadian control. Flower opening in *Cereus* and individual flowers of *Oenothera* is repeated at the same time even when the plants are kept in darkness.

In addition to flower parts, other structures may exhibit nastic movements during reproduction. The flower stalk of the poppy *Papaver rhoe* initially exhibits extreme epinasty in the part subtending the closed flower bud. It is curved vertically down by differential growth that is more rapid in its upper sector, forming a hook (as in dark-grown etiolated seedlings of dicotyledonous plants). At anthesis (flower opening) the differential growth is reversed, and hyponastic movement results in an erect flower. Poppy flowers respond to rain by very rapid curvature of the peduncle downward and loss of petal turgor, enclosing the flower and ensuring pollen is not lost.

Gymnosperms do not produce flowers; rather, they have their fertile organs arranged in a tight spiral, making a cone. Microsporophylls, each carrying two microsporangia (pollen sacs), are arranged in relatively short "male" cones. Megasporophylls, each eventually carrying an ovule (mature megasporangium with egg cell), are paper-thin bracts that are arranged singly in axils of tightly packed, thick woody scales forming the cone. However, despite their massive appearance and woody nature, the bracts unfold to allow access for pollen to the ovule (pollination), after which they refold into a lightly packed cone, leaving room for development of the embryo. Finally, these bracts unfold once more, maximally and irreversibly this time, to release the mature seed. These are all endogenously controlled, nastic movements associated with reproductive and seed development.

External Signals

In contrast to the autonomous, nastic movements, tropic movements are stimulated by exogenous directional (vectorial) or unilateral signals.

Direction of these movements is tightly coupled to direction of the signal by the physical location and orientation of sensory receptors. These sensors may be remote from the motor tissue or located within the cells of the motor itself. Positive and negative tropisms describe movements toward and away from the source of the signal, respectively. Gravitropic movements are guided by perception of the gravity vector in specialized sensory cells and anisotropic response by motor cells (see Chapter 3, Gravity-Guided Movements). Phototropic movements are guided by differential interception of unilateral light, or by interception of directional (more or less collimated) light at an oblique angle (see Chapter 5, Light-Driven Movements). Phototropic movements of leaves, flowers, or inflorescences, whose direction changes throughout the day with the changing position of the sun, are heliotropic—"sun-tracking" or "solar tracking" (see Chapter 6, Tracking the Sun; and Photo 1). Stems of many climbing shoots, as well as leaves and other foliar organs metamorphosed into tendrils, respond to friction when they encounter mechanical support and wind themselves into tight spirals around the support. These responses are often called thigmotropic, although it is not touch itself but friction that generates the movement.

Many plant movements are controlled by both endogenous and exogenous signals. Most importantly, movements under the control of the circadian clock are modified to a precise 24-hour diurnal rhythm by the daily cycle of light and darkness. This "setting" of the clock is crucial for plant function. In many species, unfolding (and refolding) of flowers synchronizes floral availability, with the optimal periods for pollination determined by activity of the pollinating agents or by the best time for scattering of the pollen, which requires the opening of pollen sacs and exposure of the stigma. Diurnal movements, so-called when they are externally modified circadian movements, are synchronized, rephased repeatedly by diurnal environmental signals such as light or temperature transitions. Movements that are synchronized by a dark/light transition are photonastic, whereas reverse movements that are synchronized by the opposite, light/dark transition are skotonastic. In nature, these transitions rephase the circadian folding and unfolding of leaves, opening and closing of stomata, and opening and closing of flowers and inflorescences. Many leaves and flowers and inflorescences fold around nightfall, which led to the term "sleep," or nyctinastic movements.

Temperature fluctuations control thermonastic movements. *Tulipa* (Liliaceae) and *Crocus* (Iridaceae) flowers open in response to a small increase in ambient temperature and close in response to lowering temperature (by as little as 1°C to 3°C). Adaxial tissue from the perianth of *Tulipa*

Photo 1. Leaves of lupine spreading out to catch light and carbon dioxide for photosynthesis.

exhibits an abrupt increase in length in response to a 10°C increase in temperature. Abaxial tissue from the perianth exhibits a similar increase in response to a 10°C decrease in temperature.

Chemical signals also may be external signals for plant movements. For instance, pollen tubes moving toward their objective ovule are guided by chemical cues from the synergid cells guarding the egg cell, their target within the ovule (Fig. 3). Where ovules are protected within an ovary, movement of the pollen tube is guided through the style by chemicals produced by cells lining the pathway through the style. These movements are chemotropic. Roots are known to grow toward water, a hydrotropic movement. Movements of stomatal guard cells, on the other hand, are nastic, hence the responses of stomatal guard cells to substomatal carbon dioxide are chemonastic.

Growth of stems and roots is generally directed by gravity; stems grow up (away from the gravity vector, with negative gravitropism) and roots grow down (toward the gravity vector, with positive gravitropism). Branch roots and stems may grow at various angles with respect to gravity, and are described as diagravitropic. Some plant organs, such as shoot apices of climbing vines, apices of young, etiolated seedlings, and tendrils, exhibit spiral autonomous movements, described as circumnutation movements. In roots, a similar behavior can be observed, for example, in *Arabidopsis* (Brassicaceae) seedlings grown on the surface of agar in sterile culture, and the behavior is called root waving. These movements are caused by

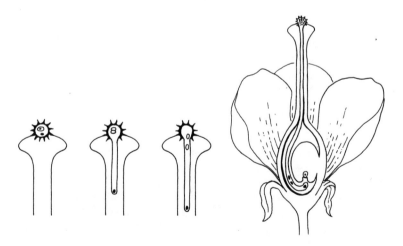

Fig. 3.   Pollen grains germinate on the stigma of a flower; when a physiological match has been found, the pollen tube grows down through the style, homing in on the egg in the ovule by sensing chemotropic signals.

Pollen Tube

All seed plants employ the single-cell motors of their pollen tubes for conveying male gametes (sperm cells) to fertilize their heavily protected female gametes (egg cells). The motor is located at the tip of the pollen tube, which is guided with precision toward its target. This is no simple task, especially in flowers where the egg cell, contained in an embryo sac, is wrapped with protective layers of maternal tissue to form an ovule. Ovules are housed in ovaries, located at the base of the style through which pollen tubes must travel. When a pollen grain lands on compatible receptive tissue, the papillar cells on a flower's stigma, the pollen tube starts growing only after complementary chemical passwords have been exchanged from the pollen grain's external coating, the exine, and the papillae have established intimate contact (adhesion).

Extension of the pollen tube occurs by "tip growth." Movement of the tip of the pollen tube along the style occurs between cells of a transmitting tissue, or along the glandular inner surface of a hollow tube in the interior of the style, guided by appropriate secretions from the cells along this pathway. At the tip of the pollen tube, many vesicles carry cell-wall components out to its apex as well as enzymes for digesting a track through the flower's style. It is not clear what force propels the tube forward; it is unlikely to be turgor as the tip's wall is constantly forming and thus is structurally unable to develop turgor. Possibly, cytoskeletal elements 'push' the formation of the new tip forward. Some distance behind its apex, the pollen tube seals itself off from its earlier self and proceeds as a plug of walled cytoplasm down through the female tissues which nourish it as it goes. The distance traveled through the style may be quite long; for instance, in maize (*Zea mays*, Graminae) the "silk" exceeds 10 cm. Within the tissues of the style, tips of pollen tubes move forward at approximately $4 \, \mu\text{m} \, \text{min}^{-1}$. In some species they may move as fast as $15-60 \, \mu\text{m} \, \text{h}^{-1}$, while in others, such as species of Cactaceae and Orchidaceae, fertilization may be delayed several weeks.

The pollen tubes eventually emerge into the lumen of the ovary where the task of guidance is taken over by an ovule, more specifically by the embryo sac located just below the entry port, the micropyle. Remarkably, only one of the numerous pollen tubes that invade the ovary directs its movement precisely to the micropyle. When the tip of the pollen tube has reached the embryo sac, it penetrates it and discharges its two gametes to join the egg.

activity of a subapical zone of elongation that is confined to the flank of the organ. The laterally located zone of elongation is dynamic and moves in a spiral, progressively upward direction toward the apex along the growing organ. The differential elongation along opposite flanks results in curvature. Spiral relocation of the zone of elongation transforms the curvature into a dynamic spiral movement of the apex as it is propelled into its environment.

In certain adult plants and many dicot seedlings, the shoot apices exhibit nonspiral, to-and-fro movements. Such nutation movements occur by spontaneous differential elongation and curvature. The more distal, gravity-sensing part of the stem (situated beyond the curvature), when displaced, responds to the deviation from gravity, resulting in an opposite curvature. Response to the signal lasts longer than its perception, and curvature may overshoot the gravity vector in a different direction. The resulting movements are random, and can be thought of as the stem seeking a vertical posture via trial and error.

Mechanics of Movement

Underlying all plant movements, whether they are generated in growing or mature plant parts, is a "motor." For many obvious movements, such as sun-tracking and sleep movements in leaves, the motor tissue is specialized for reversible movements, and is discrete and separate from the organ that it moves. In other cases, organs move as a consequence of growth; roots and root hairs penetrate the soil in an ever-enlarging network, and the pollen tube grows down through the flower's style. In these cases, the motor tissue is the growing tissue; that is, the motor is made up of cells that are expanding irreversibly. The distinction between movements caused by growth and movements occurring in mature tissues rests in whether the cell walls extend reversibly (elastic) or irreversibly (plastic extension). All plant movements, whether in growing or mature tissue, are based on physical force exerted against the structure of the cell walls. It is the location of the motor tissues, the mechanisms by which they work, and the signals controlling them that create variation in plant movements.

Moving by Expansion

Plants are continuously moving the young, potentially active parts of their bodies, relocating them to be able to take advantage of the resources that are available in their surrounding environment. This is achieved by progressive penetration of the soil environment, spearheaded by the root tip (of the primary root, of adventitious roots, and of their branches) and the progressive penetration of the aerial environment, spearheaded by the shoot tip (of the primary shoot and its branches). The root and shoot tips are propelled through their respective environments by means of a growth-mediated motor tissue situated a short distance behind the apex,

known as the elongation zone. The movement takes place by metaboli-
cally driven elongation resulting from anisotropic growth of the cells in
this zone, restricted to the direction along the axis of the root and shoot.
The tissues composing the elongation zones are in a constant state of
flux. New cells, formed in the apical meristems in the root tip and in the
apical bud, are continuously being added to the elongation zone, while
maturing cells at the other end exhaust their capacity for elongation and
join their predecessors in the maturation zone.

CHANGING DIRECTION.   Actively growing plant organs can change
their spatial orientation by growth-mediated curvature of the elongation
zone subtending the apex. Differential rates of growth in opposite sectors
of this motor tissue change the spatial orientation of the organ that it
supports; acceleration or inhibition of cell elongation in one sector, which
is usually but not necessarily accompanied by opposite changes in the
opposing sector, result in curvature of the motor tissue.

Root hairs move similarly, although they are not multicellular but rather
are lateral projections of epidermal cells. The ability of the root hair to
move forward by tip growth enables it to follow the tortuous path between the soil particles and extract the water held there by surface tension. However, the tip of the hair must be able to direct its growth around
individual soil particles that it encounters as it moves forward. Such colli-
sions shift the entire tip-growth mechanism within the tip of the root hair
to the free side, thus bulging around the obstacle while remaining within
the water-filled interstices of the soil matrix. Pressure encountered by any
of the subcellular components associated with tip growth may be the sensor that causes this change in direction of movement.

CHANGING CONFORMATION.   Some motors employ their capacity
for performing anisotropic changes in cell dimensions to execute drastic
changes in the architecture and spatial organization of the plant or specific
portions of it. At the cell level, the paired stomatal guard cells provide an
ubiquitous example of a single-cell motor that operates by conformational
changes. The specific and highly anisotropic structure and thickening of
the cells' walls transform changes in their volume into changes in their
conformation, leading to opening or closing of the aperture between them.

Leaves of certain plants, particularly among the grasses (monocotyledonous plants), provide an example of a multicellular motor that creates
extensive changes in leaf conformation. The major veins along the entire
lamina of a grass leaf are all parallel to each other. Each is equipped with
a motor running along both flanks of the vein, consisting of a single

straight row of very large, bulliform (bubble-shaped) cells. When grass leaves experience drought, anisotropic contraction of the bulliform cells causes the leaf lamina to roll up along its length. The movement is reversed when water becomes more available to the plant; the bulliform cells inflate and flatten the leaf. In a similar way, young grass leaves are formed in a tight scroll; they unfold as they emerge from their surrounding sheath and are exposed to light.

Moving by Contraction

Some plants are capable of moving their massive and bulky storage organs, such as bulbs or corms, including their apical bud, through the soil matrix to optimize their location within the underworld (see Chapter 4, The Underground World). These employ specialized contractile roots, all of adventitious origin (arising from the stem), for moving along linear paths (translation movements). The upper part of such a root characteristically metamorphoses into an unbranched, thick, fleshy organ that does not participate in absorption of water or minerals. After it reaches a predetermined length, it reverts and continues growing as a typical slender, branched absorbing root. Concurrently, the fleshy part starts to undergo extensive contraction along its axis. Anchored by the branched part at its distal end, the contracting part pulls the shoot above it down through the soil. The developmental pattern of such a root is, from its start, strikingly unlike that of the ordinary absorbing root. The radial expansion of the fleshy part of the root results from massive growth of parenchyma cells in its cortex. This process is progressive; as the root thickens, it gradually forces aside the surrounding soil, thereby creating a tunnel. This space is subsequently utilized by the plant as a conduit for moving its storage organ, avoiding mechanical and frictional resistance of the soil (Fig. 4).

Plants in some categories exhibit movements of shoots that are associated with change in priorities. Plant roots are highly conservative in structure, form, and function. There are notable exceptions where roots have undergone extensive structural modifications for highly specific tasks, such as contractile roots, storage roots, prop roots (mechanical support of the trunk), pneumatophores (roots for ventilation of submerged root systems), and aerial roots (epiphytes). In contrast, shoots exhibit an inexhaustible variety of architecture and structure. Most of these are structural adaptations associated with specific functions, such as storage of water (stems and leaves of succulents) or food (rhizomes, bulbs, corms, and tubers—all of which are modified stems), territorial invasion (runners and stolons), mechanical support (prop roots), provision of high protein diet (carnivo-

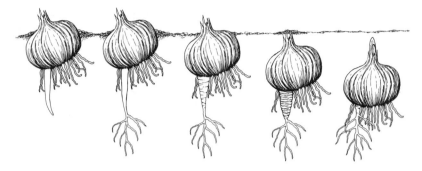

Fig. 4. A *Gladiolus* (Iridaceae) contractile root anchors in the soil then contracts, drawing the attached bulb down into the cavity behind it.

rous plants), and parasitism (see Chapter 7, Prey-Driven Movements). Clearly, the priorities in such shoots for allocating resources differ from those of the vast majority of shoots.

In their never-ending quest for sunlight, plant shoots continue growing inexorably upward and sideways. Their mass and weight also increase continuously, as does their exposure to the shearing forces of wind. Plant shoots support the weight of their aerial shoots by investing a large proportion of their resources in tissues devoted uniquely to mechanical support against the vertical forces of gravity on the one hand and the shear forces of wind on the other. Certain plants have evolved alternative strategies for supporting the weight of their shoot at a far lower expenditure of precious resources, while at the same time reducing the weight that must be supported. They do this by downgrading the priority of massive investment of resources in mechanical supporting tissues, replacing it by a priority for a much more economical investment in motor organs that do the same job. These motors specialize in locating external supports and using them with high efficiency, thus making it superfluous for the plant to manufacture its own mechanically supportive tissues. The motors characteristically are parts of the shoot that transformed into the highly specialized organs of climbing plants, by means of which they locate suitable extraneous support and become firmly and permanently attached to it.

In most cases, these organs are long, thread-like, slender, and highly flexible organs that locate external support by performing extensive circumnutations. Where its cells elongate, the organ becomes convex, and its apex moves in a continuous circular or elliptical orbit along an inverted cone, thereby constantly probing the aerial environment for support. The space explored in each cycle depends on the distance between the apex and the moving zone of elongation. In the hop (*Humulus*, Cannabaceae), the

space explored may reach a diameter of 50 cm. In *Hoya carnosa* (Asclepia-daceae), it may extend to 150 cm. Once the motor locates a suitable support, it becomes attached to it; circumnutation stops and is replaced by differential elongation that results in formation of a coil around the support. The motor transforms from functioning as a probe into the complex function as a support. Special strategies and mechanism are subsequently employed that enhance and amplify the strength of the attachment. Plants with these capabilities fall into two general categories.

TWINING PLANTS.   In the category described as twining plants, the young, actively growing stem exhibits very long, slender internodes and itself performs circumnutation, location of a suitable support, and the all-important differential growth that results in firm attachment to the support, as in hops (*Humulus*, Cannabaceae). The stem is erect (negatively gravitropic), and the axis of circumnutation is vertical. Thus, a vertical support is the most suitable. Functionally, the support substitutes for the virtual absence of mechanical support in the stem. Once suitable support has been located, the twining stem winds itself around it in a helix, which becomes progressively more tightly coiled. This transformation is obviously the result of a thigmotropic response (a directional response to unilateral touch), resulting in differential changes in length along the external flank (axial expansion) and the flank in touch with the support (axial contraction). The latter senses the presence of a suitable support, and the transduction chain extends not only across the stem but also for considerable distances along it. During the circumnutation phase, the stem is uncoiled, and any flank may establish contact with the support. It follows that the sensory mechanism for the coiling response is evenly distributed around the circumference of the stem.

TENDRILS.   The tendril category includes all species whose motor organs are designated as tendrils (Photo 2). These are metamorphosed branches (single in *Bryonia* [Cucurbitaceae] and *Passiflora* [Passifloraceae]; branched in the grape vine *Vitis vinifera* [Vitaceae]) or leaves, as in *Lathyrus aphaca* (Fabaceae). The midvein is all that remains of the lamina, and it functions as a tendril. The task of photosynthesis is relegated to the two prominent stipules. For example, in the bean—*Phaseolus* spp. (Fabaceae)—the leaf lamina is divided into lobes that progressively metamorphose into tendrils in the younger leaves; in the pinnate leaf of the pea—*Pisum* spp. (Fabaceae)—the distal leaflets metamorphose into tendrils. In contrast to twining shoots, tendrils start their existence precoiled as a torsion spring (as in mechanical watches), but they straighten out before starting cir-

cumnutation. This establishes a dorsiventral dissimilarity between the previously concave and previously convex flanks of the tendril.

In most if not all tendrils, it is only the previously concave flank that exhibits a thigmonastic response, commonly maximal to its more distal part. However, excitation of the opposite convex flank, which by itself elicits no response, may nevertheless inhibit simultaneous excitation of

Photo 2. Tendril of *Passiflora.*

its opposite concave flank, which suggests the presence of sensors in both flanks. The response starts within approximately 2 minutes by axial contraction of the cells along the initially concave flank and simultaneous expansion along the opposite, initially convex flank. Soon after, both flanks elongate rapidly but to a lesser extent along the concave flank, resulting in progressive coiling.

Once contact had been established with a suitable support, the thigmonastic response causes the tendril to wind itself tightly several times around the support in a helical coil, anchoring it firmly to the support. The coiling also increases the contact area between the sensitive part of the tendril and its supports, thus enhancing further coiling. The response is then transmitted basipetally, causing the basal parts of the tendril to form an intermediate, free-standing helical coil, frequently exhibiting an "inversion point" where the direction of coiling is reversed, thus equalizing the torsion.

In any event, the substantive formation of the coils below the support act as a spring that actually pulls the attached plant closer to the support. Whereas twining stems are inherently vertical, tendrils originate as lateral appendages of the erect shoot. Therefore, they are randomly oriented in space, and any support that can accommodate them is suitable, whatever its orientation. The extensive contraction of the coils below the support draws the plant forcibly toward the support, up, down, or sideways.

The response of a tendril to thigmonastic excitation is rapid (0.5–35.0 minutes after excitation). The thigmonastic sensory system is apparently highly selective with respect to the exciting agent. It will not be excited by a smooth, slippery object (such as a stick covered with gelatin), or by a powerful liquid jet. In the Cucurbitaceae, such as *Luffa cylindrica*, the receptor cells are visible as epidermal protuberances. The architecture of the protuberances of sensory cells differs in different species. They are distributed differentially, much more abundant on the inherently convex flank and particularly in the vicinity of the tip. The entire epidermis of the tendril is interconnected by means of plasmodesmata that traverse the anticlinal cell walls through pit fields and integrate all the protoplasts into a giant symplast. This accounts for the rapid and far-reaching transmission of the thigmonastic stimulus.

Posthumous Changes in Spatial Organization

Movements that change conformation can take place after the organ, structure, or plant has died. The motor for these movements is hygroscopic, based on a close association between two neighboring tissues with different mechanical structure. One of these tissues is highly aniso-tropic, with elongated cells and the cellulose microfibrils in their walls organized in parallel. Changes in hydration, from humidity or liquid water in the environment, result in changes in the dimension transverse to the axis of these cells. Structural differences between the two component tissues are expressed by widely divergent hygroscopic responses. The differences may be quantitative; in which case, the dominant component acts as the dynamic motor while its neighbor serves as an anchor. Their joint action results in reversible curvature. Alternatively, the hygroscopic potential of both components may be similar but oriented at an angle to each other. Both act as motors that enable the entire tissue to coil reversibly as a spring.

In arid and semiarid habitats, posthumous movements act in controlling seed dispersal (and germination) according to changes in atmospheric vapor availability or occurrence of significant rainfall or substantial dew (see Chapter 8, Liberation Movements). *Plantago coronopus* (Plantaginaceae) is a dwarf desert annual that exhibits a rosette of leaves and a number of plagiotropic inflorescence branches (peduncles). As they dehydrate during senescence, the latter curve progressively down to become tightly appressed to the soil surface. The abaxial (lower) sector of the peduncle acts as the motor, anchored on the more passive adaxial (upper) sector. Decrease and increase in the hydration of the cellulose microfibrils in their wall cause the entire abaxial sector to contract, or expand along the axis of the peduncle. In contrast, structural organization of the opposite, adaxial sector is mostly random, with limited capacity for contraction. *Asteriscus pygmaeus* (Asteraceae), a similar species, exhibits similar posthumous movements, but as it dehydrates its peduncles curve upward into tight coils (Fig. 5). In this species, the roles

Fig. 5. Energy stored in the drying cell walls of the desert dwarf Asteraceae *Asteriscus pygmaeus* fruit is released suddenly and forcefully upon hydration, flinging the seeds out and away from their parent plant.

of the abaxial and adaxial sectors of the peduncle are reversed. Contraction takes place in the adaxial sector. During fruit maturation, a motor located at the base of the calyx moves the sepals to fold upward and enclose the fruit. In these plants, the conformational change is reversed, and the contracted tissues expand during the following "effective" rainfall as they are rehydrated. The seeds are exposed, and some are detached and dispersed by the raindrops. The entire process is fully reversible and may be repeated in subsequent cycles of a dry and rainy season, resulting in repeated dispersals and attempts at germination and establishment.

Salsola kali (Chenopodiaceae) and some of its relatives are annual plants that inhabit saline deserts. Its branches are obliquely oriented (plagiotropic). As the plant senesces and dies, its branches dry out progressively. The tissues connecting the shoot to the taproot are weakened by disintegration, allowing the shoot to detach, after which it is rolled over long distances by winds, dispersing its seeds along the way. Similar tumbleweeds occur in different families: *Gundelia tournefortii* (Asteraceae), *Cachrys gravicarpa* and *Ferula biverticillata* (Apiaceae), and *Allium schubertii* (Liliaceae).

Adaptive Value of Movements

Characteristically, all motorized movements exhibit adaptive values. These adaptations may be self-evident, but some are relatively subtle and open to speculation. Movements of apices of primary shoots and roots under the guidance of the gravity vector have a clear adaptive value. The same applies to movements of the apical part of young shoots and of mature leaves under the guidance of the light vector. Posthumous hygroscopic movements have a distinct adaptive value for coinciding seed dispersal with periods of available water and for conserving seeds within the plant in the dry period.

Motor organs of certain plants perform negative phototropic movements, by which they locate a suitable vertical support to climb. The ivy *Hedera helix* locates adjacent vertical support (such as walls or tree trunks) by negative phototropic movement of its actively growing subapical regions of the stem. It then becomes attached to the support by producing on its shaded flank numerous short, stiff adventitious roots in the form of a brush; these roots penetrate crevices in the support and establish themselves there permanently and firmly. The wild vine *Parthenocissus veitchii* (Vitaceae) and its relatives carry short metamorphosed branches, each branching into a number of short extensions terminated by discs. The branches and the discs exhibit negative phototropism, growing toward a suitable neighboring vertical support. When the discs establish contact with the support, they start secreting a very potent adhesive, by which they become permanently and firmly attached to the support.

Motors that move the stomatal guard cells control the width of the aperture between them. Thus, they have full control over the import and export of gases through the epidermis, between the photosynthetic tissues and the external environment, and they are of obvious adaptive value to photosynthetic gas exchange and water conservation. Movements of root hairs into the interstices of the soil matrix result in a huge increase in their capacity for absorbing the nutritive resources of the soil solution. Movements of pollen tubes serve as the instrument of fertilization in all seed plants. Leaf movements of carnivorous plants provide them with a rich diet of amino acids and other factors of high nutritional value.

CHAPTER TWO

Motors

PLANTS ARE IMMOBILE, rooted in place during most of their life, but they are constantly in motion, restless. Some move their leaves, some their flowers, others twist their whole bodies. Roots penetrate soil, and roots and stems drag and place underground organs at optimal physiological depths. Movements take place within individual cells, by individual, free-living cells and their colonies as well as by discrete parts and organs of multicellular plants. All of these movements take place by means of motors, powered by a variety of forces.

Specialized structural, architectural, and functional features of plant cells bear direct relevance to the motors by which such movements take place. Single-cell motors act in movements of stomatal guard cells, root hairs, and pollen tubes. Multicell motors operate to move entire leaves or leaflets, roots and other subterranean plant parts, apical buds and their subtending cluster of developing leaves, flowers or inflorescences, fruits, or entire shoots. Most movements are driven by turgor (hydrostatic) pressure within the motor tissues. Movements are called growth-mediated movements when they are driven by turgor but occur in immature cells still capable of irreversible extension of their cell walls, cell volume increase, and growth. They are designated turgor-mediated movements when they occur in mature cells that undergo reversible, elastic extension of their cell walls.

Plant movements employ features that are universal and nonspecific. One of these features, the structure of the plant and its cells, plays a crucial role in the mechanisms by which plants move. Another feature is the critical role played by a variety of forces, including hydrostatic pressure, in the dynamics of plant movements.

Motor Cells and Tissues

Most categories of plant movements are driven by multicellular motors. Prominent among these is the pulvinus, a tissue generally found at the base of a leaf or petiole which acts as a joint, bending the leaf to different positions (Fig. 6). Multicellular motors occupy a distinct position with respect to the portion of the plant that they are assigned to move. As a rule, in multicellular motors the distal part of the motor is firmly attached to the base of the moving part. This allows the motor to direct the movement of the freely suspended portion. Most but not all multicellular motors are approximately cylindrical, with a central core of vascular tissue. Their motor tissue is arranged either in two opposite, multilayered sectors on either side of the vascular core, or as a complete peripheral, multilayer sheath. Differential changes in cell volume in opposite sectors of the motor tissue are expressed as differences in the axial dimension.

Pulvinus

Mature leaves equipped with a pulvinus may increase or optimize their harvesting of light each day by changing the spatial orientation of their lamina. The pulvinus functions as a crane, consisting of a hydraulic motor within a flexible "pivot" that connects a rigid stationary "post" (petiole or rachis) to a movable, rigid "boom" (midvein of the lamina, rachis, rachilla). The motor operates by simultaneously generating a turgor differential across the pulvinus; this produces sufficient torque to displace the mass of the laminar boom over considerable angles (which may exceed 90°) relative to the stationary rachis (the post). These changes in volume of the pulvinar motor tissue, and the resulting changes in its curvature, are fully reversible and repeatable throughout the active life of the leaf.

The motor cells in the pulvinus exhibit structural specializations. The bulk modulus of elasticity of the pulvinar motor tissue is low because the anisotropic changes in dimensions of their cells enable the swelling and shrinking of cells to accommodate large changes in cell volume in response to relatively smaller changes in turgor pressure. Because changes in turgor pressure involve energy-dependent processes of transport across membranes, this feature contributes to increased efficiency in converting osmotic work into cell expansion. The orientation of the cellulose microfibrils in walls of motor cells is transverse to the pulvinar axis, constraining volume changes of pulvinar motor cells along the pulvinar axis. In addition, the epidermal sleeve presents mechanical

(continued)

constraints to radial expansion of the pulvinus, while transverse folds of the epidermis along the pulvinar axis contribute to reducing mechanical resistance to axial expansion/contraction of the subtending motor tissue. Veins coalesce in the pulvinus to form a flexible but nonextensible core (accommodating changes of ~180°), allowing the pulvinus to change its radius of curvature without changing its length. Leaf movements frequently involve torsional rotation (of the petiole or pulvinus). The structural features of such torsions are far from clear. One possibility is that the cells along the lateral sectors of the (cylindrical) organ are structurally constrained to expand or contract preferentially along their diagonal axis.

The pulvinus exhibits precise physiological coordination in the operation of its motor. In the trifoliate leaf of bean, opposite but equivalent volume changes take place simultaneously in opposite sectors of the pulvinus. Solutes and water are lost from cells on the contracting side and taken up by cells on the expanding side, driven by membrane transport processes regulating cell osmotic concentration, turgor, and water potential. These processes are fueled by the respiratory and synthetic activity of the motor cells, which contain large, prominent nuclei; an abundance of well-developed mitochondria with tightly packed cristae; well-developed chloroplasts with prominent grana stacks; numerous polysomes; and an extensive endoplasmic reticulum (rough and smooth). All are indicative of the high capacity of the motor cells for generation of metabolic energy.

Motors that move young, immature parts of the plant are dynamic. These growth-mediated motors are in a constant state of flux, as they move to keep up with the movement that they create. They act by means of irreversible increase in volume along their axis as their cells undergo elongation. Cells at the base of the motor that have completed their elongation are no longer part of the motor. Fresh, newly formed cells join the motor at its apex. These immature cells are just starting elongation and replace the departing, mature cells. The cell population is in a constant state of flux, but its number, and therefore also the size of the motor, remain unchanged. Motors that move apices of roots and shoots are typical examples.

Motors that move fully mature parts of the plant are static. These multicellular motors are composed of highly organized, fully mature, specialized tissues that are permanently situated at the base of the organ they move. They act by means of fully reversible changes in volume (expansion or contraction) along their axis. Reversibility is achieved by high elasticity and other properties of the motor cell walls. Increase or decrease in cell volume takes place by increase or decrease in the hydrostatic pressure

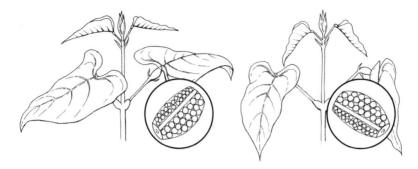

Fig. 6. The pulvinus is specialized tissue acting as a joint which, by differential cell swelling and shrinking, changes leaf position.

(turgor pressure) acting on the cell walls from within the vacuole. Such motors are turgor mediated, as in the pulvini that move the lamina of most leaves and leaflets.

Some of the most fundamental plant movements involve single cells, and their unicellular motor is part of their specialized structure. Unicellular, growth-mediated motors operate in actively growing cells that move by anisotropic growth at their tip, as in tip growth. Root hairs are common examples of such growth-mediated movements, as are pollen tubes, which exhibit the universal example of such movements in all seed plants by which the male gamete moves and is guided toward union and fertilization of a genetically receptive female gamete.

Unicellular motors operating in mature single cells are turgor mediated. Movement is the result of anisotropic changes in volume caused by anisotropic thickening of the cell walls. Such movements change the conformation (spatial architecture) of the cell. The ubiquitous stomatal guard cells are an example of such unicellular motors. Their movements change the stomatal aperture, determining the exchange of gases (carbon dioxide, water vapor, and oxygen) between the photosynthetic tissues of the plant and the external environment (Fig. 7).

Stomata

The most widespread turgor-mediated motor is the stomatal guard cell, a single-celled motor. Paired guard cells sit parallel to each other in epidermal tissue; they are in contact only at their ends and form the stomatal pore be-

(continued)

tween them. Cuticular waxes covering epidermal cells render their surface virtually impermeable to gases such that the guard cells' movements opening and closing the stomata control the exchange of carbon dioxide, water vapor, and oxygen between the ambient atmosphere and underlying tissue.

Most plants have kidney-shaped guard cells, but grasses and other monocots have more specialized, dumbbell-shaped guard cells. Changes in cell volume (expansion and contraction) and conformation of the guard cells are fully reversible for the lifetime of the cells. Deformation takes place in guard cells as a result of the specialized, nonuniform thickening of their walls. Movements of guard cells are determined solely by the anisotropic structure of their walls and are therefore nastic. They move only in the plane of the neighboring epidermis.

Stomatal opening takes place when turgor pressure in the guard cell exceeds that in its adjacent subsidiary cell. These two cells work in concert, pressing against each other. In *Tradescantia virginiana* (Commelinaceae) pressure in the guard cells increases progressively from ~0.05 to ~0.7 MPa as stomata open to a 26-μm aperture, while the (calculated) difference in solute potential between the guard cells and their neighboring subsidiary cells increases from ~0.15 to ~0.8 MPa. The volume of the subsidiary cell is several times greater than that of the guard cell. Thus, the transfer of a small fraction of solutes from the subsidiary to guard cell results in a much greater proportional increase in solute potential in the latter than the decrease of solute potential of the former.

Changes in solute potential correspond with changes in vacuolar potassium (K^+). In *Allium cepa* (Liliaceae), changes in K^+ are matched perfectly by changes in chlorine (Cl^-). In other species, such as *Commelina communis* (Commelinaceae) and *Vicia faba* (Fabaceae), the counter anion in these salts does not appear to be Cl^-, or any other inorganic ion, but an organic acid. Malic acid and other acids of the tricarboxylic acid cycle are prime candidates in these species. Solute uptake is driven by proton and electrical gradients set up by the proton ATPase (pump) on the plasma membrane.

Under constant environmental conditions, changes in guard cell volume exhibit an autonomous circadian rhythm. The diurnal cycle between light and darkness rephases the cycle to 24 hours; stomata are open during the light phase and closed during the dark phase. In light, opening is driven primarily by blue wavelengths interacting with the photoreceptor phototropin to enhance proton pump activity. Guard cells also respond to changes in substomatal carbon dioxide: in darkness, respiration increases its concentration, closing the stomata; in light, photosynthesis takes up the gas, causing stomata to open. The sensors for carbon dioxide are located within the guard cells.

Stomata close during water deficit, a critical behavior for preserving plant hydration. Closing is enhanced by the stress hormone ABA (abscisic acid), which is supplied by the chloroplasts of the underlying mesophyll cells and from other plant organs (primarily roots) in response to drought. Guard cells have

(continued)

specific receptors for ABA which signal inhibition of the proton pump (H+–ATPase) and subsequent loss of solutes. Interestingly, each environmental cue influencing swelling or shrinking of guard cells must have the opposite effect on the subsidiary cells. Questions about how this motor, the guard cell, is regulated may be answered in part by investigating its partner, the subsidiary cell.

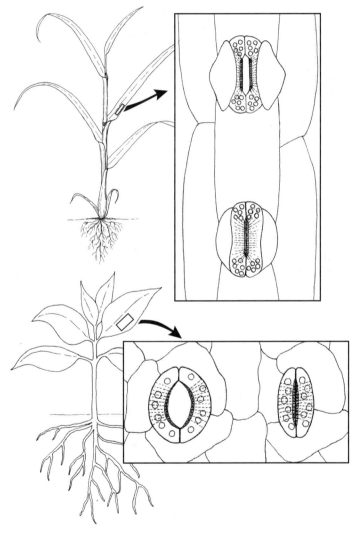

Fig. 7.   Stomata are pores between paired guard cells that control passage of gases in and out of the plant. Grasses (*top*) have dumbbell-shaped guard cells; broad-leaved plants (*bottom*) have sausage-shaped guard cells.

Growth-Mediated Motors

Differential rates of growth in opposite sectors of the motor change the spatial orientation of the organ that it supports. Acceleration or inhibition of cell elongation in one sector of the support (which is usually but not necessarily accompanied by opposite changes in the opposite sector) result in curvature of the motor. Young, actively growing organs may also change their spatial configuration by differential growth of their own tissues. Cell growth is a direct, irreversible consequence of increased extensibility of the cell wall, resulting in irreversible expansion of cell walls by osmotic uptake of water, balanced by osmotic adjustment of solutes.

Certain single-celled plant organs that are universal and essential for critical functions of the plant move through their environment by means of a growth-mediated motor situated at their very apices. Such tip growth takes place by guiding all newly produced cell wall components to the tip of the cell where synthesis of new cell walls takes place. The root hair uses this kind of movement to guide itself between the soil particles in search of the resources of the soil solution held there by capillary forces. The fern's prothallium faces similar problems. A lesser-known example is the elater of the liverwort sporangium, which uses a motor operated by surface tension. The most ubiquitous of these tip-growing motor organs in all seed plants is the pollen tube, which guides the male gamete to a union with its female partner exclusively within the same species or in a closely related one.

Root Hair

Root hairs extend the surface area of roots by a thousandfold in the soil. These single-celled organs navigate soil obstacles by tip growth, in a fashion similar to the pollen tube. Root hairs develop from specialized, short-lived epidermal cells called trichoblasts, which are located in the majority of plants just above the zone of elongation of all roots. As their name suggests, they are tubular, slender (5–17 μm diameter) extensions of trichoblasts that protrude perpendicularly into the environment and may reach a length exceeding several millimeters.

A root hair starts growing when the trichoblast acidifies (pH ~4.5) a spot in its external wall where expansin proteins are located. Acid activates the ex-

(continued)

pansins, which catalyze wall loosening, allowing the wall to bulge outward. Sustained growth of the root hair takes place by continuous increase in the volume of the (central) vacuole (estimated at 50 fl per minute). This requires constant input of osmotically active solutes (ions) to maintain osmotic potential at a level that is sufficient to build up turgor. Tip growth proceeds quite rapidly (~1 μm s^{-1}), supplied with newly synthesized cell-wall components transported rapidly to the growing apex by vigorous cytoplasmic streaming, probably involving contractile protein elements of the cytoskeleton. Microtubules are organized in parallel along the axis of the root hair and direct the traffic of cell-wall components. Tip growth is closely supervised by the nucleus that migrates into the tube and stays close to the site of activity at its tip. Elongation is associated with a fivefold increase in $[Ca^{2+}]_{cyt}$ from ~200 to ~1,000 nmol focused on the site of most rapid growth. The direction in which the tip of the root hair moves is determined by the location of this gradient. If the gradient is focused directly at the tip, the hair will grow straight; if the gradient moves in response to external signals around to the side of the hair's tip, it will turn and grow in a new direction.

Dimensions of the root hair provide a great increase in the contact area between the root and its soil environment, increasing the efficiency of uptake of water and nutrients. Assuming the surface area of a mature trichoblast to be $20 \times 20\ \mu$m, and the diameter and length of a mature root hair to be 10 and 1,000 μm, respectively, then the surface area of the hair is greater by a factor of about ×75.

Development of root hairs is tightly coupled to availability of nutrient ions in the soil solution in their environment. The correlation with nutrient concentration is an inverse one. The lower the concentration, the greater the density of root hairs and their length. At 1 mmol phosphorous, root hairs grow five times faster than at 1,000 mmol. The number of root hairs forming cell files increases 8 to 12 times, and three times as many of these cells develop into root hairs. However, in the soil medium, hairs can survive only as long as they maintain contact with available soil solution, held by capillarity within the larger pores of the soil. A hair's life ends prematurely when the soil solution is exhausted, after which point the root hair withers and dies; its function is taken over by newly formed hairs closer to the young root's apex.

Growth is irreversible. Therefore, maturation of cells comprising the motor tissue puts an end to its capacity for growth-mediated movements. The growth-mediated motor is therefore characterized by inherent obsolescence. However, adjacent younger tissues or new, younger cells beyond the motor may then take over the motor function. In this case, the motor is dynamic and simply moves forward (acropetally). Growth-mediated curvature can only be reversed by compensatory growth in the opposite (concave) sector, but only as long as the motor tissue retains its capacity for growth. After this point, curvature becomes permanent.

Orientation of young shoot organs and their aggregates (cotyledons, the seedling plumule, young expanding leaves, the apical bud and its cluster of young leaves, entire flowers or their parts, or entire inflorescences or their parts) changes with curvature of the stem or other subtending tissue supporting them. The direction of light and gravity are the dominant environmental factors in the aerial environment, and both therefore play a major role in controlling these movements. These environmental signals are perceived in different sensory organs, but both operate by means of the same motor. Therefore, light-guided movements may deflect the organ from its response to the gravity vector, in which case the responses to light and gravity are integrated.

A latent potential for growth-mediated movement may be retained in specific mature tissues of certain plants. The base of the leaf-sheath at the grass internode (also known as a pulvinus) retains a latent capacity for growth-mediated curvature in response to gravistimulation. For instance, if the stem collapses by "lodging" as a result of heavy rain, snow, or strong wind, one or more pulvini exhibit negative geotropic curvature, thus restoring the shoot above into an erect orientation. The resurrected vertical shoot regains its capacity for normal function (Fig. 8).

Turgor-Mediated Motors

In contrast to the growth-mediated motor, the turgor-mediated motor is composed of fully differentiated, mature cells. This kind of motor is stationary, permanently situated. Its structure is clearly distinguishable and delimited from its neighboring cells and tissues. Its movements do not involve growth and are fully reversible and repeatable throughout maturity. The turgor-mediated motor has a lifetime warranty.

Changes in turgor are the result of transmembrane transport of osmotically active solutes, primarily ions. Such transport is powered by bioelectricity (electric power generated by biological systems). Turgor-mediated movements are less costly than growth-mediated ones in that the resources needed to build the cell walls and tissue connections are only needed once. Both types of movements invest metabolic energy in the transport of osmotically active solutes across cell membranes. Changes in volume take place as water follows the transport of solutes. Turgor-mediated movements invest additional metabolic energy in the membrane turnover that is a consequence of the extensive changes in volume, while

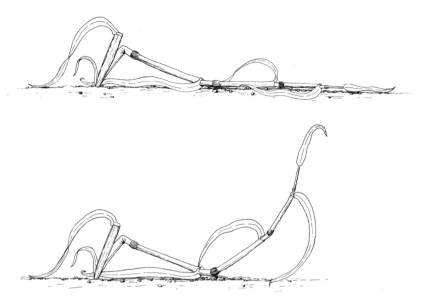

Fig. 8. Specialized motor tissue at the base of monocotyledonous (grass) leaves can respond to gravity, reorienting leaves upward after flooding has caused them to fall over or "lodge."

growth-mediated movements invest more metabolic energy as well as resources in the biosynthesis of new cell walls and cytoplasmic components. With the exception of stomatal guard cells, these motors are multicellular.

Of all the plant organs, only the leaf has evolved specialized tissues that are structurally and specifically adapted to facilitate its rapid, turgor-mediated reorientation, or changes in its conformation. The mature leaf lamina of many plants may exhibit extensive conformational movements in response to changes in leaf water availability. The lamina of mature leaves of dicotyledonous plants may fold along their midrib in response water loss. The lamina of a monocotyledonous plant may roll up along its length, changing from a flat, expanded configuration to a compact, scrolled configuration in response to similar stress. This response reduces exposure of the lamina to solar radiation and dry air. Furthermore, changing the configuration of a flat, long lamina to a tightly rolled cylinder is particularly efficient in reducing interception of solar radiation because the angle of incidence becomes progressively more acute toward the sides. Both types of leaves revert to the flat configuration when the stress diminishes sufficiently. These movements take place by means of a turgor-mediated motor consisting of specialized inflated epidermal cells (bulliform cells) situated in a single parallel row along their major vein(s) with their bases anchored on both sides to the neighboring line of epidermal cells. The motor acts by means of anisotropic and fully reversible changes in volume of the bulliform cells (contraction and expansion transverse to the midvein).

Leaves of some insectivorous plants employ a turgor-mediated motor in their traps that operates instantaneously to close the trap on their prey when it stimulates their tactile organs by contact with them. In *Dionaea*, the inflated, highly turgescent cells of the motor keep the trap open; when stimulated, cells of the motor collapse, causing the trap to close. This action is reversible and turgor mediated.

Origins of Forces

Changes in the dimensions of motor cells and tissues are the result of specific forces generated uniquely by one of two physical processes. Both of these processes are based on the special properties of water. One process is based on electrostatic attraction of water molecules to each other and to other charged molecules. The other process is based on the ability of water molecules to move across cell membranes much more rapidly than most other molecules dissolved in water. The former is responsible for movements involving senescing or dead plant tissues and organs. The latter is responsible for movements involving living cells, tissues, and organs (Fig. 9).

Cell Structure

The structure of plant cells is a direct consequence of their autotrophic nutrition. They synthesize all their organic constituents and nutrition from external resources: carbon dioxide from the atmosphere, and water and nutrients from the soil solution. Synthesis of macromolecular structure, enzymes, vitamins, and other metabolites is driven by chemical energy converted via photosynthesis from sunlight. Resources for the process of photosynthesis are in the environment outside the cell and must be absorbed into its cytoplasmic interior before they can be processed. To do this, resources must cross the cell's external boundary, the plasma membrane. The volume of the plant external to this membrane is termed the apoplast, and

(continued)

the combined interior volume is the symplast. From a structural viewpoint, the external surface area of the cell, the plasma membrane, is a vital limiting factor for all its activity. Plants have solved the problem of acquiring necessary nutrients from dilute sources by adapting several specific structural features:

1. Increase the interaction between the cell and its environment by increasing surface area.

2. Minimize resistance to diffusion of gases (CO_2, O_2, and H_2O) into the cytoplasm by limiting the cytoplasm to a thin layer in direct contact with the external boundaries of the cell. A vacuole, a vesicle containing water and solutes but not protein, fills the interior of the cell.

3. Maintain the rigidity of this potentially collapsible structure by enclosing the cytoplasm within a flexible cell wall with limited extensibility. Plant cell walls are complex carbohydrate structures based on cellulose microfibrils of high tensile strength that are embedded in a matrix of hemicelluloses and pectic compounds. Osmotically active solutes within the vacuole create a positive hydrostatic turgor pressure that stretches the enclosing cell wall. The architecture of most but not all plant cells is permanently fixed by the architecture of their walls.

4. Increase interception of solar radiation by organizing the chloroplasts in a thin layer, equally spaced within the cytoplasm and parallel to the cell surface. This arrangement maximizes the interception of diffuse light, characteristic of spongy mesophyll. Light penetrating the leaf is refracted and reflected repeatedly from the cell walls of the extensive intercellular spaces within that tissue. Photosynthetic tissue of the leaf that is directly exposed to directional sunlight is generally composed of cylindrical cells, arranged parallel to each other and to the direction of sunlight (palisade mesophyll). Their chloroplasts are aligned along the longitudinal walls, thus providing equal access to incoming sunlight.

5. Enhance traffic between the cytoplasm and its environment by cytoplasmic streaming (cyclosis), the continuous movement of the cytoplasmic layer and its metabolic as well as its light-harvesting components (chloroplasts) around the boundaries of the cell.

6. Connect domains of cells by cytoplasmic connections called plasmodesmata, tubules that allow transport of nutrients and signaling molecules from cell to cell via the symplast, minimizing the energy needed for crossing out and back in through the plasma membrane.

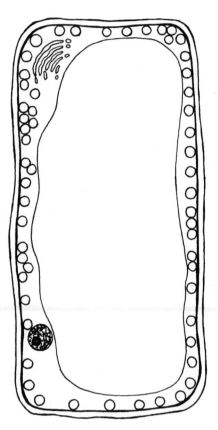

Fig. 9. A cross-section of a palisade mesophyll cell shows the large central vacuole and the arrangement of chloroplasts around the cell's periphery. Plant cells are bounded by a plasma membrane and rigid, cellulosic cell wall.

Water Molecules Acting Together

Covalent bonding between an oxygen atom and two hydrogen atoms forms the molecule of water. In covalent bonding, one partner contributes an electron to a common orbital around both partners. The single electron of each hydrogen atom shares an orbital with the oxygen atom, which leaves its nucleus positively charged. The two hydrogen atoms are attached to the oxygen atom asymmetrically, which leaves the nucleus of the oxygen atom negatively charged. The positive charge of the hydrogen nucleus (proton) is attracted to negatively charged nuclei, such as the oxygen atom of another water molecule or similar atoms (such as nitrogen) of other molecules. This attraction results in a weak electrostatic bond called the hydrogen bond. Thermal agitation of the molecules generates kinetic energy that may disrupt this bond, making its lifetime dependent on temperature. This accounts for the change of state of water from liquid to solid (ice) as temperature falls below the freezing point, or from liquid to vapor as temperature rises above the boiling point. Ions and other charged water-soluble compounds form hydrogen bonds with water.

Hydrogen bonding is common between water molecules and many giant molecules (macromolecules) or crystal aggregates. Macromolecules are polymers assembled from smaller, identical building blocks (monomers). When such monomers bear electric charges, these charges are closely and evenly distributed on the surface of the macromolecule. Similarly, aggregates of minerals are composed of crystals in which identical structural elements are repeated at precise distances. In a macromolecule or crystal that is electrically charged, the charges are closely and evenly distributed on the surface. Water molecules that form hydrogen bonds with charged particles (macromolecules or crystals) are closely packed and highly organized in parallel with each other (bound water). Other water molecules form hydrogen bonds with the first layer. These bonds are also highly organized, but to a lesser degree than the first layer, and they are not as parallel. This is repeated in successive layers of water molecules, whose orderliness, organization, and packing decrease progressively. The density of the bound water is highest at the surface and decreases progressively in the more distant, less tightly packed, less orderly layers, eventually reaching the density of free water. The layer of bound water increases the effective dimensions of the molecule to which it is bound. Thus, the effective volume of such macromolecules or crystal aggregates increases when they come in contact with water (liquid or vapor). This is expressed as imbibition, or swelling. Opposite changes (shrinkage) take place when the bound water becomes depleted, such as by evaporation.

Imbibition and shrinkage may be isotropic or anisotropic. Properties of matter may be the same in all directions, a condition described as isotropic. When properties of matter differ in specific directions, the condition is described as anisotropic. Imbibition of fluid in isotropic matter results in isotropic swelling, equal in all directions. Imbibition of fluid in anisotropic matter results in selective swelling in predetermined directions. The same principles hold in the case of shrinkage by water loss.

The cellulose matrix of the cell wall plays a major role in movements powered by joint action of water molecules. Cellulose, the principal component of plant cell walls, is highly anisotropic as a result of its structure.

The structural unit of cellulose is glucan, a very long macromolecule. Glucan is a polymer composed of a very large number (6,000–12,000) of identical molecules (monomers) of the six-carbon sugar (hexose) β-D-glucose. The molecule of this sugar is organized in a six-member ring of five carbon atoms (numbers 1 to 5) and an oxygen atom (a heterocyclic compound). Carbon atom six, with its hydroxyl group, forms a side chain. A covalent (glycoside) bond binds the rings (glucose residues) to each other by condensation between the hydroxyl groups of carbons 1 and 4 to form a very long chain. Because of the β configuration, neighboring rings face in opposite directions, and the angle of the oxygen bridge joining them is opposite alternately. Consequently, the entire glucan molecule is virtually straight, zigzagging back and forth. In this configuration, the hydroxyl group of carbon 3 comes close to the neighboring oxygen bridge that connects adjacent sugar rings and forms a hydrogen bond between them. This adds rigidity to the otherwise flexible straight glucan chain.

Glucan molecules may aggregate into supermolecules. This happens when neighboring chains lie tightly packed in parallel to each other, sufficiently close to form hydrogen bonds; in this case, bonds form between the glycosidic oxygen bridge linking two neighboring rings in one glucan chain and the hydroxyl group of carbon 6 in a glucose molecule in the parallel chain. Such bonding repeats itself, zipping together parallel lengths of glucan chains into flat sheets. Sheets of glucan chains also come to lie closely parallel to each other, allowing hydrogen bonds to form between them. As a result, the parallel sheets reinforce each other, similar to plywood. In this supramolecular region, the distances between neighboring glucose residues along the glucan chain are identical, and the same is true between neighboring chains and between neighboring sheets. This forms the cellulose micelle, a crystal-like (paracrystalline) structure, joining as many as 100 glucan chains (cross section 2×4 nm). Individual glucan chains extend beyond the micelle into several micelles, joining them to each other. In this way, several micelles (approximately 20) participate to form a loose, stable, three-dimensional network of cellulose microfibrils (consisting of 30 to 75 glucan chains, cross section approximately 3×5 nm), the basic unit of the plant cell wall.

In many cases, numerous microfibrils coalesce (more or less parallel to each other) to form macrofibrils, visible by microscope. Most of the hydroxyl groups within the micelle take part in hydrogen bonds and are therefore not available to form such bonds with water molecules. In contrast, hydroxyl groups and oxygen in the intermicellar portions of the cellulose microfibril are exposed and free to attract water molecules and imbibe. Therefore, the typical X-ray diffraction is retained in imbibed cellulose microfibrils. However, this imbibition can only take place transversely to the direction of the glucan chains and is therefore highly anisotropic. Furthermore, portions or regions of the cell wall in which the cellulose microfibrils are deposited more or less parallel to each other will also exhibit anisotropic imbibition. Taking up or losing water by such walls will cause them to change dimensions anisotropically in a preferential direction, transverse to the axis of the microfibrils.

Anisotropic imbibition and contraction play a critical role in posthumous movements of senescing or dead parts of a plant. If the cell walls of adjoining or neighboring tissues exhibit differences in the orientation of their anisotropic microfibrils, considerable tension may develop between them when they swell or contract. As a result, they may separate, twist, or curl into separate spirals.

ROLE OF SURFACE TENSION.   Water molecules in the liquid state exhibit random movements by forming random, temporary hydrogen bonds with their neighbors. The forces of attraction acting on them neutralize each other. Taken together, these forces are considerable and account for the phenomenon of cohesion. Liquid water can withstand tremendous tension (negative pressure) before the attraction between the water molecules is broken (as may happen in the water-conducting vessels of tall trees).

The situation is different at the interface between the water and air. Water molecules at this surface are subject only to forces of attraction by other molecules along this interface and beneath it. This attraction results in the phenomenon of surface tension, which is a force that is characterized by a tendency to shrink the surface as much as possible (as in spherical water drops and bubbles).

The interface between water and air may be intersected by a solid barrier. What happens at this intersection depends on the physical relationship between the water and solid. When the solid surface is hydrophilic, it attracts the water molecules by a process of adhesion. This creates a force that acts along the interface and increases the surface tension. As a result, the water creeps up the solid surface, against gravity, forming a concave lip (the degree of curvature is expressed by the contact angle α). In contrast, when the solid surface repels water, it is hydrophobic. Surface tension is reduced, and the contact area of the water with the solid becomes convex ($\alpha < 90°$).

Surface tension and adhesion of water act to produce capillarity. Water in a vertical capillary tube (where r is the radius) rises. The force causing this rise acts along the periphery and is a function of the circumference: $F_{up} = (2\pi r)T\cos\alpha$ (where T is the absolute temperature). At equilibrium, the water column reaches a height h, and this force is counterbalanced by the opposite force of gravity and is a function of the volume of water $F_{down} = (\pi r^2 h)\partial g$ (where ∂ is the density of the water, and g is the acceleration due to gravity). Therefore, the height of the water column is inversely proportional to the radius of the capillary tube, $h = 2T\cos\alpha / r\partial g$, and it changes in inverse proportion to the change in radius. The pressure within the water column is negative (less than atmospheric) in proportion to its height. Therefore, the tension of water in the capillary is also inversely proportional to the radius. In other words, the tension of water in the capillary becomes greater as the diameter of the capillary becomes smaller.

INTERACTION WITH THE CELL WALL.   Most of the wall of mature plant cells consists of a network of cellulose microfibrils. Water is bound to the cellulose matrix and fills the spaces between the microfibrils. These interfibrillar spaces form a network of irregularly shaped microcapillaries. Water evaporating from the interface with air increases the surface tension and therefore also the tension of water in the capillaries. In living cells, the increase in tension causes water to move from the vacuole to the cell wall along the gradient in water potential, but in dead cells there is no vacuole to supply water. The increase in tension inside the microcapillaries causes them to shrink, and their narrowing causes an additional increase in tension. If the lumen of the dead cell is initially filled with water, this water will be taken up into the shrinking capillaries, and the entire cell now shrinks. If the walls of such a cell are thin, they will collapse. Thicker walls resist collapse, but they may become deformed elastically. As water is lost from the lumen, the remaining water will come under increasing tension, causing the cells to contract elastically. This additional tension eventually overcomes the opposite force of surface tension. At this point, the capillary will empty of water, and air will force itself into the cell lumen. The tension within the lumen is released immediately, and the cell expands abruptly due to the elastic strain in its walls. Certain movements of senescing or dead plant parts employ such capillary devices as an aid to dispersal of spores.

Water Molecules Acting Individually

Water molecules move (by means of their temperature-dependent kinetic energy) by diffusion in random directions within the space available to them. The velocity is a function of temperature. Solute molecules do so as well, independently of the movement of water molecules. Net movement by diffusion is along the concentration gradient, from regions of high concentration to those with lower concentration. Solute molecules displace water molecules in the solution. Therefore, net movement of water molecules also takes place along their concentration gradient from regions of their higher concentration (low solute concentration) to regions with lower concentration (higher solute concentration). The relative concentration of water molecules is known as the solute potential Ψ_s (with pure water as the reference point). Therefore, the solute potential of a solution is always negative.

Membranes in general, and biological membranes in particular, exhibit selective permeability to molecules (and ions) moving through them by diffusion. When two aqueous solutions (of the same or different solutes) are separated by a selectively permeable membrane, the rapid movement of water molecules through the membrane will take place along the gradient in solute potential (from the more dilute to the more concentrated solution). This unilateral movement of water molecules by diffusion will continue until the much slower diffusion of solute molecules eventually equalizes the concentration on both sides of the membrane. Alternatively, if pressure is applied to the receiving solution (higher solute concentration, lower solute potential) the density of the water increases, resulting in higher concentration of water and elevated solute potential. Eventually, equilibrium is reached, and net movement of water by diffusion ends. In terms of potential, the negative solute potential has been diminished by addition of the positive (hydrostatic) pressure potential Ψ_p. Their sum comprises the water potential Ψ_w of the solution: $\Psi_w = \Psi_p + \Psi_s$. At equilibrium, the hydrostatic (turgor) pressure is balanced precisely by the opposite wall pressure. The process by which diffusion of water molecules takes place across a selectively permeable membrane along the gradient in water potential is osmosis (solute potential = osmotic potential).

Turgor-mediated movements take place by differential changes in volume (expansion or contraction) of fully mature cells as a specific result of transport of osmotically active solutes, followed by water from or into their vacuole. Walls of mature cells can only expand or contract elastically. Therefore, volume changes in such cells depend on the elastic modulus of their wall and are associated with corresponding changes in their turgor pressure. Expanding/contracting cells may also change their shape to enable large changes in cell volume in response to relatively smaller changes in turgor pressure.

Transport of osmotically active solutes, accompanied by water, across the membranes of the cells is an integral and essential part of the motor in turgor-mediated movements as well as in growth-mediated movements. In the cell, uptake of water by osmosis builds up hydrostatic (turgor) pressure against the equivalent, opposite pressure of the wall. At equilibrium, the cell water potential is balanced precisely by the solute potential in the apoplast: $\Psi_w^{cell} = \Psi_p + \Psi_s^{cell} - \Psi_s^{apoplast}$. In the absence of osmotically active solutes in the apoplast, $\Psi_s^{apoplast} = 0$.

Cell growth disrupts the equilibrium. It is initiated by an increase in the extensibility of the cell walls, in many tissues controlled by the hormone auxin. This reduces wall pressure and lowers the (negative) water potential of the cell. Water diffuses into the cell, increasing its turgor, which leads to an increase in volume (growth). The influx of water into the cell dilutes the osmotically active solutes. This shortfall is compensated by active uptake from the apoplast. Such uptake maintains the gradient in solute concentration between the protoplast and the apoplast. Any of these processes can be considered a prime target for the signals that drive growth-mediated movements.

Growth and consequently growth-mediated movement are powered by turgor, as is turgor-mediated movement. The only difference between growth-mediated and turgor-mediated movements lies in the extensibility versus the elasticity of the cell walls of the motor tissue. Extensible cell walls that are stretched elastically beyond a certain well-defined "yield threshold" are incapable of reverting to their original dimensions; elastic cell walls have no such threshold, which makes the growth-mediated movements irreversible and the turgor-mediated movements fully reversible.

Growth and Turgor Regulation

The force that drives growth-mediated motors is irreversible cell expansion. Growth is driven by turgor pressure generated osmotically by uptake of water into the cell vacuole, which is generally accompanied by synthesis of intracellular components and organelles. As the cell expands, an increase in surface area inevitably involves synthesis of the cell wall, plasma membrane, and vacuole membrane. Plant cells normally increase 10-fold or more in volume after they are formed. Extension of roots and shoots, expansion of leaves, elongation of root hairs and pollen tubes, gravitropic and phototropic bending, and even the snapping shut of the Venus' flytrap (*Dionaea*) are ac-

(continued)

complished by pressure-driven, irreversible expansion of cells. Movement of stomatal guard cells and pulvini is turgor-driven, reversible expansion and contraction of cells.

The rate of cell expansion depends on two factors: the force exerted against the walls due to turgor pressure, and the extensibility of the cell walls. Both are under cellular regulation. Turgor pressure results from water passively entering the cell's vacuole as it flows down a water potential gradient ($\delta\Psi$) from outside to inside the cell. The cell's water potential (Ψ_w) is lowered by either accumulation of solutes or relaxation of the cell wall. To the extent that the cell can take up or generate osmotic solutes and loosen the cell wall, it will be able to attract water inside and generate turgor to drive cell expansion.

Solute uptake occurs across the membranes via transport proteins, including pumps, transporters, and channels. The plasma membrane is energized by an active proton pump; this enzyme utilizes energy released from ATP, and it exports protons (H^+) into the cell wall space (apoplast). Proton pump activity hyperpolarizes the membrane (cytoplasmic side negative) to approximately -150 or $-200\,mV$, and sets up a pH gradient (cytoplasmic side, $pH = 7.0$; apoplastic side, $pH = 5.0$). Both of these gradients support transport of osmotic solutes into the cell. Many substances, including anions (Cl^-), sugars, and amino acids, are cotransported into the cell with protons, utilizing the favorable proton gradient. Cations (K^+) flow into the cell down the electropotential gradient. Membrane activity therefore can be viewed as the ultimate motor for growth-mediated and turgor-mediated movements in plants, as these fluxes drive solute accumulation, water uptake, and generation of turgor.

For growth to occur, turgor pressure must succeed in irreversibly extending the cell wall. The primary wall is made up of cellulose microfibrils embedded in a matrix made of hemicelluloses, pectins, and proteins. An increase in the wall surface area requires that the wall be loosened, extended, and refurbished. Wall loosening occurs by controlled release of hydrogen bonds between cellulose and hemicelluloses, carried out by acid activation of the protein expansin. Other wall-loosening enzymes address matrix bonds. Extension occurs if turgor pressure exceeds the wall's yield threshold. Wall synthesis restores structure and prepares the wall for further loosening. Without wall loosening, cells may expand reversibly but will not grow.

The rates of solute accumulation and wall loosening are under regulation by the cell. Hormones such as auxin and the photoreceptor phototropin stimulate cell expansion by increasing proton pump activity; in developing cells, this causes wall loosening as well. The stress hormone abscisic acid inhibits cell expansion and causes stomatal closure by reducing pump activity and stimulating ion efflux, leading to turgor loss.

Contractile Proteins

Mature plant cells are characterized by their relatively large volume, most of which is taken up by (usually) one central membrane-bound vacuole filled with water and solutes. The living cytoplasm forms a thin layer bound by a tonoplast around the vacuole and by the plasmalemma (cell membrane) adjacent to the cell wall. The numerous and different functional components of the cytoplasm are distributed in this extensive thin layer. Interaction between these components is essential for cell functioning, and such interaction is enhanced by the continuous mixing of these components by a cytoplasmic motor.

THE UBIQUITOUS MIXER. Intracellular motors generate the force to operate peripheral cytoplasmic streaming (cyclosis) within the cytosol, primarily along the cell walls but also through strands that traverse the vacuole. Cytoplasmic streaming is visible, despite the transparency of the cytoplasm itself, because of the numerous intracellular organelles and other cytoplasmic inclusions of various sizes such as the nucleus, chloroplasts, mitochondria, other metabolic complexes, inert cytoplasmic inclusions, and vesicles that are carried along within the streaming cytoplasm along the cell walls. Some inclusions are carried passively; others are attached to components of the motor itself. Streaming increases their contact with the intracellular environment and its various metabolic units as well as with the extracellular environment and its resources.

The motor responsible for intracellular movements is part of the cytoplasm in the form of microfilaments. Microfilaments are thin and thread-like, and they are capable of contracting and expanding their length reversibly by use of metabolic energy (adenosine 5'-triphosphate, ATP). Reversible changes in the length of microfilaments provide the motors for cytoplasmic streaming.

Microfilaments are composed of two separate contractile proteins: actin and myosin. Actin consists of subunits of G-actin, a globular protein, arranged in two parallel chains that are twisted around each other to form a helical fiber (5–7 nm in diameter). Myosin, the other protein that takes part in the structure of microfilaments, is made up principally of two long fibrous protein chains coiled around each other in a tight helix (13–22 nm diameter); the chains separate at one end to form a compact double head.

Microfilaments are the universal mechanism for moving and mixing the cytoplasm in plant cells. Actin molecules are polar and embedded in the immobile peripheral layer of the cytoplasm, anchored to the cytoskeleton.

Myosin interacts with actin by forming reversible cross-bridges with it. In animal cells, the interaction between actin and myosin is responsible for muscular contraction, the animal's essential element of movement.

Cytoplasmic streaming depends on repeated cycles of such interaction. During contraction, the head of the myosin thread slides along the actin filament, from one subunit to its neighbor, dragging its tail along. Each such movement uses up ATP. Therefore, the rate of streaming is a function of the rate of making and breaking these cross-bridges, namely, on the turnover rate of myosin-ATPase. In darkness, the ATP supply is provided by respiratory oxidative phosphorylation. In light, cells equipped with chloroplasts supply ATP provided by photophosphorylation. Cytoplasmic streaming is accelerated by increasing temperature (5°–30°) as well as by high-irradiance blue light. The action spectrum peaks at 450 nm and 366 nm, with a minimum at 400 nm. This suggests a flavin as the photoreceptor. Use of linearly polarized blue light supports the concept that the photoreceptors are localized in the immobile peripheral layer of the cytoplasm in a dichroic orientation, parallel to the long axis of the cell.

THE UBIQUITOUS MOVER. Intracellular movement to specific destinations takes place by means of microtubules. For example, microtubules are the basis for moving chromosomes to their precise destination during nuclear and cell division. The basic units of the motor take the form of microtubules: elongated, flexible, rod-like tubes (20–30 nm diameter) that are usually associated in bundles, organized in highly structured organelles. Bundles of microtubules are the prominent structural and functional element in spindle fibers and asters. Spindle fibers are attached to the individual chromosomes that they move (pushing or pulling) during karyokinesis (nuclear division) through the viscous cytoplasmic medium. The spindle fibers are able to change their length reversibly, in this way moving what they are attached to. Microtubules are composed of subunits of the globular protein tubulin, thirteen of which make up the circumference of the tubule. They can change their length by adding or subtracting subunits (polymerization and depolymerization).

Microtubules also act as intracellular guides that delimit well-defined pathways for moving newly formed macromolecular components of the cell to their destination in the course of cytokinesis (cell division). They form the dynamic structure of the phragmoplast that provides channels for guiding with high precision the movement of vesicles formed by the Golgi apparatus to the plane of cell division. These vesicles are loaded with the macromolecular components of the cell plate and, subsequently, the middle lamella that separates the daughter cells. Microtubules also guide macromolecular components of the cell wall to their destination and determine the orientation of the newly formed cellulose microfibrils within the wall. Microtubules are major components of the cytoskeleton that act as the scaffolding anchoring the actin molecules and supporting the internal organization of the cell that determines its shape. By changing their length and distribution, they change the topography of the cell.

THE PROPELLERS.   Unicells and their colonies move through their aqueous environment by means of microtubules. Microtubules are the fundamental structural and functional element of flagella and cilia, the universal organelles by which single-celled animals and plants (such as unicellular algae or their colonies) or reproductive cells (such as gametes or spores) swim through an aqueous environment. Epithelial cells of certain animal tissues are equipped with cilia, which they use to stir the aqueous medium in their vicinity and enhance take up of their nutrients and oxygen.

Flagella and cilia are identical in their function as well as in their structure. These thread-like organelles protrude from the cell surface into the liquid medium, well beyond the cell surface, in the form of flexible threads (0.25 μm diameter), enclosed in a sheath of the cell membrane. They differ only in length and in their distribution. Flagella and are borne singly or in very small numbers in a restricted location in the cell. Cilia are numerous and arranged closely together in large patches or fields. Flagella are much longer (100–200 μm) than cilia (2–10 μm). Flagella (in general) perform either whip-like, rhythmic movements or spiral movements, by which they propel the cell either forward or backward through the liquid medium. Cilia move back and forth rhythmically and in unison, usually in the same direction.

The microtubules in a flagellum or cilium make up its axonema, its flexible core. The axonema is composed of nine pairs of microtubules, fused in parallel (doublets) and arranged in a circle around a central inner sheath containing a pair of microtubules. One microtubule in each peripheral doublet extends an arm to its neighbor in a clockwise direction. Another arm extends as a spoke to the inner sheath. The arms are made of a protein (dynein). The tip of the dynein arm is attached to a tubulin subunit in the neighboring microtubule, and it can move to the next tubulin subunit along that microtubule. This movement requires metabolic energy (ATP) and leads to displacement of neighboring microtubule doublets along each other. The result is curvature of the axonema. Such movements take place rhythmically in succession in neighboring doublets and result in a whip-like or spiral movement of the flagellum or cilium.

Movement of flagella and cilia by rhythmic curvature results from differential changes in length in opposite sides of the axonema. This demonstrates a fundamental principle of movement: movement by curvature invariably takes place by differential changes in dimensions on opposite sides of the structure involved. Movements of cells, tissues, and organs of higher plants follow the same principle.

CHAPTER THREE

Gravity-Guided Movements

IRDS AND HUMANS moving over large distances depend on universal, permanent, absolute, and stable reference systems to guide their travel to its destination with absolute precision. Among such systems are the stars, the magnetic field of the earth, and the solar transit. Plants depend on the gravitational field of the earth, which is an equally universal, stable reference system for navigation and for guidance in orienting their display of organs. Gravity is the most permanent, constant force among all environmental forces acting on plants. This explains why all plants have evolved sensory mechanisms that detect the direction of the gravity vector with uncanny accuracy, using it as a compass. All plants and their organs move in directions that are tightly coupled to the direction of the gravity vector. These are designated gravitropic movements, and they are guided by gravity in one, two, and three dimensions.

The primary shoot and root (as well as most adventitious roots) are the only plant organs that exhibit orthotropic (vertical) movement throughout their existence. Branches of the primary root and shoot, and their higher order branches are plagiotropic, that is, they characteristically grow at an angle with the organ from which they originate. The lamina and petiole of leaves are dorsiventral and have one plane of bilateral symmetry. When they are displaced from their stable relation with the gravity vector along their plane of symmetry, they respond by curvature in that plane. When leaves are similarly displaced along an oblique with their plane of symmetry, their petiole undergoes torsion so that their lamina reverts to its original, stable orientation. As a result, branches of the shoot and root have an upper and lower side, but do not exhibit a clear structural distinction between them. Therefore, designating the upper and lower sectors as dorsal or ventral is a matter of convention based on a functional discrimination.

Rhizomes are subterranean branches of the shoot that characteristically grow horizontally through the soil. The horizontal, two-dimensional movement is termed diagravitropic, and it has sometimes been ascribed to equilibrium between a negative gravitropic force and an opposing epinastic force. For example, rotating a rhizome 180° around its axis results in upward curvature of its apex, suggesting that both forces are now acting together rather than in opposition.

Functional Consequences of Gravitropism

Orthogravitropic, plagiogravitropic, and diagravitropic responses differ not only in the number of dimensions they operate in, but more importantly in their functional consequences. The primary root and shoot perform one-dimensional movements exclusively. The one-dimensional orthogravitropic response directs the growth of the primary shoot, from seedling emergence, vertically upward toward the light, and growth of the primary root vertically downward toward the moisture and dissolved minerals that have percolated downward by the same gravitational force. However, harvesting these resources is restricted to a relatively short segment, a short distance behind the apex of the primary root. In the root, the root hairs harvest the water and minerals within their reach (several millimeters) then wither and die. Newly formed root hairs that develop acropetally take over. Similarly, in the shoot, the young, fully expanded leaves are the most active in harvesting light and carbon dioxide and using them for photosynthesis. Older leaves senesce and die, and newly expanded leaves take over. Thus, linear movement can provide only a linear (constant) supply of resources. Furthermore, as the harvesting apparatus continues to move forward, transport becomes progressively more difficult.

The situation changes dramatically by the process of branching and the universal three-dimensional plagiotropic response of branch shoots and roots (Fig. 10). Each root and shoot produces a number of branches that in turn also produce branches of progressively higher orders. The spatial orientation of each branch differs from that of its contemporaries, and that of each generation of branches differs from that of its predecessors. Their harvesting capacity and the resulting supply of resources increase exponentially. Moreover, the distances of transport remain manageable because the plant structure is more compact. Each branch is oriented in a different direction, which makes exploitation of the resources more economical.

In contrast, the primary function of the two-dimensional diagravitropic portions of the specialized branch roots and shoots is neither nutritional nor photosynthetic, respectively. Their movement in two dimensions allows the plant to invade and subsequently control extensive territory around itself, eventually producing erect, leafy descendants from buds (axillary in the case of shoots, or adventitious in the case of roots).

Fig. 10. Both the shoot and root systems of plants produce multiple orders of branches, only some of which grow vertically up or down. The rest are plagiotropic, growing at some angle with respect to the gravity vector.

Gravity is a permanent directional signal that controls the direction of plant growth. However, other environmental variables may act as signals for modifying or changing the direction of growth more optimally. Growth-mediated movements can only take place in actively growing tissues. Thus, if a plant, or one of its organs, is exposed simultaneously to more than one directional signal or gradients in environmental variables, the responses are integrated within the same tissue. Therefore, these responses share at least some of the terminal steps in the signaling system leading to movement.

Light interacts with gravitropism in two different modes. One is by its presence, irrespective of its direction (tonic effect). Seedlings of *Zea mays* cv. Merit (Poaceae) produce roots that exhibit a diagravitropic (horizontal) growth in darkness. In the presence of red light, their response changes to positive gravitropic, and they grow vertically down. Phytochrome is the photoreceptor for this effect (see Chapter 5, Light-Driven Movements). In other systems, red light (also acting via phytochrome) may enhance or reduce the gravitropic response. Hypocotyls of dark-grown seedlings of *Arabidopsis thaliana* (Brassicacea) exhibit normal, positive gravitropism in their roots and negative gravitropism in their hypocotyls. In the presence of red light, the hypocotyls grow at an angle to the gravity vector. The active form of phytochrome (P_{fr}) presumably modulates their gravitropic response, but neither the mechanism nor the significance of this phenomenon is known. The other way light may interact with gravitropism is by its phototropic effect. In shoots, simultaneous exposure to unilateral light from one direction and to a gravity signal from a different direction results in an intermediate response: curvature whose magnitude and direction represent the resultant of the direction and magnitude of these two vectors.

In roots, response to gravity may be modified by a directional response to a gradient in water potential (hydrotropism) as well as a response to unilateral mechanical stimulation (thigmotropism). In both cases, the evidence suggests that these signals act on the same sensory mechanism as gravitropism via the statoliths (discussed later in this chapter). These gravity detectors are degraded on the water-stressed side of the columella or on its mechanically stimulated side, resulting in reduced responsiveness to a gravity signal. These interactions enable the gravitropic root to navigate around obstacles and toward soil moisture.

One-Dimensional, Vertical Movement

The primary root and primary shoot are the two organs of the young plant that invariably exhibit one-dimensional, vertical movement. The primary root grows vertically down in the direction of the gravity vector, exhibiting a positive gravitropic response. The primary shoot grows vertically up in the direction opposite to the gravity vector, exhibiting a negative gravitropic response. The gravitropic responses of the primary root and shoot are the most fundamental adaptations of plants to the terrestrial environment. The absorbing zone of the root produces root hairs that penetrate the pores of the soil matrix, absorbing the nutrient solution held in them by capillarity and eventually exhausting these supplies. These root hairs soon die. However, as the root tip moves through the soil, it produces fresh root hairs behind it to invade and exploit fresh soil resources.

As the apical bud at the top of the stem moves through hitherto uninhabited reaches of the aerial environment, it brings its expanding leaves within reach of a more abundant, readily available supply of photosynthetically active radiation (PAR; range = 400–700 nm) from sunlight, which they may harvest to power their photosynthetic activity. As they expand and harvest more of the available PAR, leaves face increasing competition from neighbors, all of which are expanding as well. Their vertical movement enhances their capacity to compete for light with the new leaves of their neighbors. Lower story leaves frequently exist under a deficiency in PAR, obliged to carry out maintenance photosynthesis; some may die of starvation. However, as the apical bud moves into hitherto unoccupied atmosphere, it produces new leaves, free of competition, temporarily at least.

It is in this context that the universal control by gravity over the direction of movement of the primary shoot and root can be understood, because water moves down through the soil by gravitation, and the probability of maximizing the harvesting of light in an environment where all plants grow side by side increases vertically. Directional control by gravity takes place by means of a sensory mechanism that enables the root and the shoot to determine the direction of the gravity vector, coupled with a mechanism by which these organs respond to this signal.

Sensing the Gravity Vector

The sensory mechanism for gravity consists of distinctive cells called statocytes (standing cell) because they contain specialized amyloplasts called statoliths (standing rock) to describe their special role in sensing gravity. In roots, the statocytes are located in the columella, a group of special cells at the core of the root cap (calyptra). In stems and a variety of shoot organs, the statocytes are located in the starch sheath. The density of the starch grains is greater than that of the cytoplasm in which they are embedded, which is why they move through the cytoplasm under the force of gravity, clustering at the lowest part of the statocyte. Wherever they are eventually deposited, their weight and size deform the subcellular organization of either the cytoskeleton network of microtubules and microfilaments or of the endoplasmic reticulum. This is the first step in the transduction chain by which the gravity signal is perceived and transmitted to its site of action, where it expresses itself as the

Sensors

Plants sense external environmental signals as well as internal ones using a variety of sensors. For some signals, such as light and a few hormones, the sensor molecules are known. For example, blue light interacts with at least two sets of photoreceptors, cryptochromes and phototropins. The receptors for several peptide hormones have been identified, and the presumed receptors, which are at least binding proteins for other hormones including auxin, are known.

In general, sensors are the molecules that interact with signals and set in motion a signal-transduction pathway leading to cascading, amplifying responses that ultimately transduce the signal to a cellular response. For a relatively simple case, consider the action of light causing stomata to open. Blue light activates phototropin, initiating biochemical reactions that lead to stimulation of the proton pump, accumulation of osmotic solutes, increase in turgor pressure, and stomatal opening. In the more complex case of gravitropism, the sensor is understood to include interaction of statoliths (dense particles within gravity-sensing cells) with cytoplasmic elements on the bottom of the cell, thus transmitting the direction of the gravity vector to the cell. Positioning of the statoliths influences transport of the hormone auxin, such that auxin moves toward the bottom of the organ, with less to the top. This is part of the signal-transduction process; auxin presumably interacts with an auxin receptor, which in the case of roots signals reduction of growth on the lower side and enhancement of growth on the upper side, leading to bending of the root downward.

Sensors are the switches that govern the motors driving movement. Some of the signals are endogenous, emitted under circadian control, governed by developmental signals coming from other regions of the plant, and possibly co-governed by metabolic state using sugar and other metabolite levels as co-signals with hormones. The signals are often more abundant than responses to them; this is determined by the presence or absence of receptors as well as by the transduction machinery with which the receptors need to interact. Expression of genes encoding the receptors, the transduction machinery, and synthesis of communicating signals within plants are an essential control point for normal plant development and the capacity to respond appropriately to environmental signals.

response. In roots and stems, the site of action is the zone of elongation, where growth is most active. The response is opposite in roots and stems, where it takes place by imposing differential rates of axial growth (elongation) of these tissues (primarily the epidermis) in opposite sectors.

Responding to Gravity

The primary root and stem are radially symmetrical in their structure and function. Therefore, so long as the gravity signal is perceived at the base of the sensing cell (statocyte), the response to it is also symmetrical, resulting in straight growth (Fig. 11). However, when the axis of the plant is tilted at an angle to the gravity vector, the statoliths are displaced asymmetrically, to the lowest point in the cell. As a result, the response is also asymmetrical, resulting in its curvature toward the gravity vector by differential growth on the opposite sectors of the zone of elongation.

ROLE OF AUXIN AND ITS TRANSPORTERS.   Cell elongation is controlled by the plant hormone auxin. Transport of this growth regulator is close to the end of the transduction chain involved in gravitropism. Auxin is transported from the shoot apex to the root apex through the phloem. At the root apex, the direction of auxin transport is reversed:

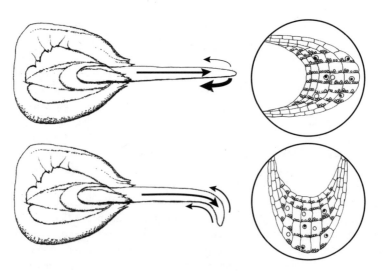

Fig. 11.   Germinating corn seeds have an elongating primary root. The enlargements detail the root cap cells which contain statoliths, starch grains that fall to the bottom of the cell, indicating which side is down.

auxin is now transported back from the root apex through the cells surrounding the phloem by means of polar (unilateral) auxin transporters embedded in the transverse membranes at the base of the cell. When the axis of the root is displaced from the gravity vector, statoliths are displaced from the base of the statocyte to the side facing the gravity vector. As a result, polar auxin efflux transporters are activated in the longitudinal cell membranes along that side and inactivated along the opposite side. Thus, the signal resulting from displacement of the statoliths to the side of the statocyte enhances the transport of auxin to the corresponding sector of the zone of elongation, while transport to the opposite sector is diminished. This results in net lateral transport across the root to the sector facing the gravity vector.

The concentration of auxin in cells in the zone of elongation determines their growth response, which exhibits a distinct optimum. In roots, the optimum concentration is very low, but the tissue concentration of auxin is supraoptimal, so increase in the concentration actually inhibits growth, and its reduction promotes growth. In stems, the optimum concentration is considerably higher, but the tissue concentration is suboptimal, so increase in the concentration promotes growth, while its reduction reduces growth. Displacement of the statoliths to a sector of the statocytes leads to enhanced transport of auxin to the corresponding sector of the zone of elongation. The signal resulting from displacement of the statoliths to one sector of the cell activates the auxin efflux transporters in the cell membrane of that sector, thus enhancing the transport of auxin to the corresponding sector of the zone of elongation and reducing transport to the opposite sector.

The primary root responds to the lateral gradient in auxin concentration by reducing elongation in the sector receiving excess auxin, and enhancing elongation in the opposite sector to which auxin transport is reduced. The ascending gradient in auxin concentration across the zone of elongation results in differential growth, causing the root to curve down in the direction of the gravity vector and to exhibit positive gravitropism. The stem of the primary shoot exhibits negative gravitropism, a diametrically opposite curvature with respect to the gravity vector. These growth-mediated curvatures reorient these organs vertically, in line with the gravity vector.

ROLE OF CYTOPLASMIC CALCIUM CONCENTRATION.   A number of elements of the transduction chain that transmits the gravity signal to its site of action have been (tentatively) identified. One such element is the concentration of the calcium ion in the cytoplasm, $[Ca^{2+}]_{cyt}$, of statocytes. Redistribution of intracellular Ca^{2+} precedes redistribution of auxin in the transduction chain. In the columella, displacement from the gravity vector and the resulting deformation of the cytoskeleton (or endoplasmic reticulum) create a gradient in $[Ca^{2+}]_{cyt}$ across the root cap. The calcium ion interacts with pectin and thus contributes to controlling the extensibility of growing cell walls. Deformation of the cytoskeleton may also open calcium channels in the cell membrane, thus activating subsequent elements of the chain. This may be mediated by the phosphoinositide pathway.

ROLE OF PROTON PUMPS. Proton (H^+) pumping across the membrane by H^+-ATPase may also be an element in signal transduction. Deviation from the gravity vector results in the transport of protons from the cytoplasm of statocytes in the columella to their apoplast, and from there to the site of response. Increase in acidity (proton concentration) is an element in the control of cell wall extensibility.

The gravitropic responses of the primary root and stem are opposite, and they also differ in their transduction pathways. Genetic mutants have been identified with normal gravitropic responses in one organ and an absence of such responses in the other. For example, the *AGR* gene encodes a protein that controls auxin efflux (and the gravitropic response) in roots but not in shoots. The gene *AUX1* is expressed as an auxin influx carrier in a single layer of statocytes in the root columella and controls the gravitropic response of the root, but not of the shoot.

Plant organs exhibit differential gravitropic responses along their plane(s) of bilateral symmetry. Dicotyledonous seedlings, such as *Cucurbita pepo* (Cucurbitaceae), *Helianthus annuus* (Asteraceae), and *Aquilegia vulgaris* (Ranunculaceae), exhibit two vertical planes of bilateral symmetry, at right angles to each other. One is along the midribs of the two cotyledons, and the other is between the cotyledons. The response to change in the gravity vector is greater along the midribs. The grass seedling has only one vertical plane of bilateral symmetry, bisecting its coleoptile, and its tropic responses (to gravity and light) are strongest along this plane. These differences are related to structural organization along the plane(s) of symmetry.

Three-Dimensional Movements

Plants branch their shoots repeatedly, increasing the surface area for harvesting electromagnetic energy of light from the sun and carbon dioxide from their aerial environment for photosynthesis. They branch their roots repeatedly, and many even produce adventitious roots and thus increase their contact area for harvesting water and nutrients from their soil environment. From a functional viewpoint, this pattern of branching allows the roots and shoots to add the element of volume to their linear exploration and exploitation of environmental resources.

Branches of a dominant primary shoot or trunk and those of a dominant taproot move in three dimensions. The prevalent spatial orientation of these branches is at a predetermined acute angle with respect to the direction of the gravity vector. These movements are designated plagiotropic. When their spatial orientation is coupled to the direction of the gravity vector, they are designated plagiogravitropic movements. Plagiotropic movements are not necessarily plagiogravitropic. Branches of the shoot and root that move in directions that are determined by the spatial orientation of the organ from which they originated are designated autotropic.

Plants use three-dimensional movement to expand into their immediately available space and its resources. Branch shoots and roots that exhibit three-dimensional (plagiotropic) movements contribute most of the biomass of the plant. They also provide the only means to exploit the volume of dilute resources in their environment. Yet very little is known about the underlying mechanism that controls these plagiotropic movements.

Formation of Branches

Branches formed by roots and shoots differ fundamentally. Meristems of branch roots develop from the pericycle of the young root at sites related to the location of elements of the primary conducting tissues, which are arranged radially and therefore exhibit equally spaced azimuth angles (Fig. 12). They emerge through the cortex and epidermis at right angles to the axis, positions they characteristically maintain for a brief period of growth before moving obliquely down. They maintain this plagiotropic orientation for some time, eventually establishing a positive gravitropic response by starting to move along the gravity vector. This pattern delays their vertical growth down, increasing the distance from their original root and the other branch roots. Competition between the roots is thus reduced, and exploitation of resources is enhanced.

Branch shoots develop from buds in the axils of leaves. Consecutive leaves are produced by the shoot along an ascending spiral at characteristic, species-specific angles and at predetermined distances from each other (phyllotaxis). As a result, branches of the shoot also exhibit equally spaced angles and intervals along the axis. Their axillary origin orients

Geometric Parameters

Canopies of trees and bushes, as well as branches in smaller herbaceous plants, are complex three-dimensional structures designed for the optimal display of leaves, their light-catching antennae. Similarly, the architecture of the root system is displayed three-dimensionally in the soil to optimize mining of nutrients and water absorption from the soil. The spatial orientation of the shoot and root branches can be described using three measurable angles:

1. Axial angle (Ax) between the branch and the axis from which it emerged

2. Azimuth angle (Az) between the plane of the axial angle and an arbitrary reference plane (e.g., north–south)

3. Angle of elevation (Ae) between the branch and the gravity vector (e.g., vertical up = 180°; vertical down = 0°; acute < 90°; obtuse > 90°)

Both Ax and Az are determined autotropically by endogenous factors. Higher order branches of the root and shoot exhibit three-dimensional movements. The Ae of such branches may differ considerably from each other and from their branch of origin, placing them outside the concept of a species-specific gravitropic set-point angle (GSA). Primary roots of germinated seedlings of maize (*Zea mays,* Poaceae) are favorite objects for studies of plagiogravitropism. When the seed is oriented vertically, the root grows down along the gravity vector. When seedlings are displaced laterally, the root curves vertically down to follow the gravity vector. However, when the seedling is displaced along its plane of symmetry, its root curves to assume an acute plagiotropic angle (<90° from the gravity vector). Light strongly enhances the downward curvature (more acute plagiotropic angle). Roots of horizontally displaced seedlings eventually reach their stable plagiotropic orientation at an angle to the gravity vector. When displaced again at this point to the precisely opposite plagiotropic angle (upper and lower sides of the root reversed), their plagiotropic angle changes, which suggests an absence of a GSA.

These results suggest that the plagiotropic angle is not a set-point angle but rather is primarily determined by graded orthotropism, involving two separate gravity-dependent controls. One of these is vectorial (dependent on the direction of the gravity vector) and induces a progressive, irreversible curvature toward a positive gravitropic orientation along the gravity vector. The other control is tonic (dependent on the progressive change in the magnitude of gravity during curvature, as well as on presence of light, not on its direction) and determines the duration, and thus the extent, of curvature toward the gravity vector (the plagiotropic angle). The role of this control is to facilitate the orthogravitropic movement, but it deteriorates progressively, stopping the orthogravitropic curvature at an inclined orientation with the gravity vector (plagiotropic angle).

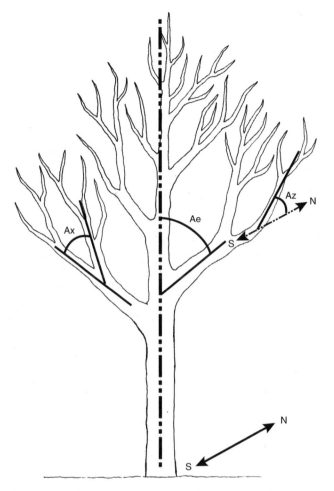

Fig. 12. A diagram of geometric parameters describes branching angles.

their initial movement obliquely upward. In contrast with roots, branch shoots may maintain this angle for extended periods of growth before exhibiting a (negative) gravitropic response by upward curvature of their actively growing, subapical part. However, characteristically this upward curvature rarely achieves the vertical. Furthermore, such curvature may be reversed by opposite curvature in their subtending, maturing part, reverting to the initial position. The duration of these stages may vary considerably in different species.

Effect of Gravity

Gravity plays a different role in the primary organs, roots and shoots, than it does in their branches. An oblique orientation that is controlled by gravity is plagiogravitropic. The angle (Ae) of a root or shoot has been designated as the gravitropic set-point angle (GSA), suggesting that it is tightly coupled with the gravity vector but may be modified by environmental variables such as light. This concept has been generalized to include the vertically growing primary root (GSA = 0°) and primary shoot (GSA = 180°), obliquely growing branch roots (GSA > 0° < 90°) and branch shoots (GSA > 90° < 180°), and horizontally growing branch roots and shoots (GSA = 90°). However, only the movements of vertically growing primary roots and shoots and of their horizontally growing counterparts are unequivocally directed by the gravity vector and are justifiably classified as orthogravitropic and diagravitropic, respectively. Classification of obliquely growing (plagiotropic) branch roots and shoots as gravitropic cannot be taken for granted unless they exhibit a response to a change in the direction of the gravity vector throughout their ontogeny. In absence of such response, their movement is more correctly defined as autoplagiotropic.

Movement of first-order branches differs in roots and shoots. An autotropic branch root must first become gravitropically competent, after which it characteristically grows obliquely down $(90° < Ae > 0°)$ before changing direction to vertically down $(Ae = 0°)$. This change in the relationship with the gravity vector takes place as part of the branch root ontogeny, despite the fact that the gravity vector remains unchanged. Displacing the branch root from its relation with the gravity vector or inverting the entire plant (by 180°) affects only the gravitropically competent part of the root and moves it to regain its prior GSA. However, first-order branch roots of young seedlings of wheat (*Triticum aestivum,* Poaceae) rotated on a horizontal clinostat to counteract gravitational force do not change their angle with respect to the axis from which they had emerged (Ax), suggesting that the initial Ax is determined autotropically. The oblique orientation may be accounted for by assuming that it is the result of two opposite forces: a positive gravitropic component and an opposite component that represses root elongation (such as auxin) and is transmitted from the source root along the upper sector. Eventually this factor is exhausted, and positive gravitropism expresses itself fully.

In contrast, the branch shoot retains its initial Ax (<90°) for extended periods as it elongates. The Ae of shoot branches may be autotropic. Primary shoots of *Tradescantia* (Commelinaceae), *Araucaria* (Araucariaceae), and *Phyllanthus* (leafflower, Phyllanthaceae) are erect, but their first-order branches maintain their original *Ae* after they have been excised. On the other hand, *Ax* of shoot branches may be determined by signals from its primary (erect) shoot. Decapitation of the primary (erect) shoot causes its first-order branches that are closest to the cut to curve upward. In other plants, inclining the primary shoot from the gravity vector, or reducing the effectiveness of this vector by means of a clinostat, diminishes the effectiveness of this signal.

In the groundnut (*Arachis hypogaea*, Fabaceae), the primary stem is erect. Its first-order branches start their development at an extremely acute *Ax*, which becomes progressively obtuse, ending up close to 90°. Second-order and higher order branches also grow out at the same axial angle. Growing the plants on an inclined clinostat modifies the gravitational effects and results in considerably more acute axial angles. Even displacing the entire plant horizontally results in upward curvature of the primary shoot, an orthogravitropic response. The first-order branches do not exhibit a similar orthogravitropic response, except in their subapical zone of elongation. However, their *Ax* stays more oblique (less obtuse), while their *Az* remains unchanged.

The branch exhibits (partial) negative gravitropism in its subapical part only during periods of active growth. This partial upward curvature may become permanent, or it may revert to the original *Ax* by compensatory growth along the upper sector during maturation of the curved part. The gravitropic competence of the lateral shoot while it continues moving at the initial *Ax* cannot be taken for granted without unequivocal evidence, such as a change in direction of movement in response to displacement from the gravity vector.

SENSORY MECHANISM FOR PLAGIOTROPISM. An oblique *Ae* (GSA) must be reconciled with the generally accepted concept of statolith displacement as part of the sensory mechanism of gravitropism. The basic difference is between the lateral root and shoot. The primary root is oriented vertically, and its sensory mechanism occupies a cluster of cells (columella) in the root cap. Displacement of the axis of the root from the gravity vector results in lateral transport of auxin across the root, resulting in downward curvature. In contrast, the lateral root is inclined, and the statoliths are inherently displaced sideways. Hypothetically, this should cause a positive gravitropic response, which evidently does not take place. This may be explained by hypothesizing that in the lateral root the sideways displacement of the statoliths is the stable point of equilibrium at which the auxin distribution remains equal in all directions, leaving the orientation of the root unchanged. Displacing the branch root from its customary angle with the gravity vector disrupts the equilibrium, causing curvature that restores the equilibrium.

The primary (erect) stem senses changes in the direction of the gravity vector by the displacement of statoliths in its endodermal starch sheath. In the erect, primary shoot, the statoliths are located at the base of the statocytes. Displacement from the gravity vector causes lateral displacement of the statoliths toward the inner longitudinal wall in the upper sector and toward the outer longitudinal wall in the lower sector. This results in differential signals from the opposite sectors, leading to lateral transport of auxin from the upper to the lower sector and resulting in differential growth and upward curvature. In contrast, branches of the shoot are inherently inclined to the gravity vector. The inherent equilibrium position of the statoliths is not at the bottom of the statocytes, but rather is shifted sideways: statoliths in the upper sector of the starch sheath are localized (by gravity) at the low point of the inner longitudinal wall of the statocytes, while those in the abaxial (lower) sector are localized at the low point of the outer longitudinal wall. Assuming that the branches are gravitropically competent, the differential signal results in equal growth in the upper and lower sectors and is therefore a permanent feature. As long as the branch shoot maintains this orientation, it continues growing at the endogenously determined oblique Ax (=Ae). The equilibrium breaks down by deflecting the shoot from its oblique relationship with respect to the gravity vector. Alternatively, it may alter in response to developmental signals during active growth, resulting in the upward curvature of the subapical zone of elongation.

ROLE OF OPPOSING FORCES.   An alternative hypothesis for plagiotropism in branch shoots is based on the assumption that the inclined orientation is a result of equilibrium between gravitropism and a force acting in the opposite direction. Species of *Abies* have an erect primary stem with primary branches almost perpendicular to it. Decapitation of the primary shoot results in upward curvature of the branches up to 60° to 70°. Application of auxin to the cut surface keeps the branch perpendicular to the primary stem or even curves it down (as much as 20°).

Gravitropism is positive in branch roots and negative in branch shoots, and the opposing forces are hyponastic and epinastic, respectively. Assuming that epinasty in plagiotropic shoots is due to the supply of auxin favoring the adaxial (upper) side of the shoot, the hyponasty of plagiotropic roots may be ascribed to a similar distribution pattern of auxin. In roots, auxin would be supraoptimal and inhibit elongation along the adaxial (upper) surface, resulting in an upward force, counterbalanced by the positive gravitropic force. In both cases, the apex of the axis (root tip or apical bud) from which the branch originated apparently controls the factor opposing gravitropism, probably by means of auxin. Excision of the apex of the primary root of the broad bean *Vicia faba* (Fabaceae) leads to production of orthotropic (positively gravitropic) branch roots near the cut surface instead of plagiotropic ones. Similarly, branches developing nearest the point of excision of the apical bud of *Impatiens roylei* (Balsaminaceae) exhibit upward curvature, but application of auxin to the cut surface reverses the curvature.

The equilibrium point depends on the plagiotropic angle (Ae). The vertical gravitational force, g, can be resolved into a component that acts perpendicular to the axis of the organ ($g \sin Ae$) and a second component that acts parallel to the axis ($g \cos Ae$). The latter has no gravitropic effect, but its magnitude may determine the strength of the opposing force that equilibrates with the perpendicular component (tonic effect). Deflecting a plagiotropic organ from its stable Ae has opposite consequences for the two opposing components. For instance, deflecting a plagiotropic root from its stable Ae to a more acute angle decreases $\sin Ae$, and thus the gravitational component; at the same time, it increases $\cos Ae$, thereby enhancing the opposing component. The root will move accordingly to increase its Ae to its stable orientation. Similar changes take place when a plagiotropic branch is deflected from its stable (obtuse) Ae to a more obtuse angle. Therefore, such displacement will revert to the equilibrium point at the stable Ae.

Diagravitropic, horizontally growing organs such as rhizomes respond to deflecting to an obtuse Ae less than to deflecting it by an identical angle in the opposite direction. The difference might be attributed to the component $\cos Ae$, which acts along the axis with the same magnitude but in diametrically opposite directions. This difference on polarity might be expressed in the rate of supply of auxin. The relationship with $\sin Ae$ was supported by results with *Tradescantia,* in which the primary shoot is vertical but the laterals are plagiotropic. The measured force opposing gravity was approximately twice as great in *Tradescantia viridis* (Ax of the branches 20°; $g \sin Ax \sim \frac{1}{3} g$) than in the related *Zebrina pendula* (Ax of the branches 10°; $g \sin Ax \sim \frac{1}{6} g$).

In several cases, the oblique spatial orientation of shoot branches has been attributed to equilibrium between an upward directed force— negative gravitropism—and a downward acting epinastic component. This may be demonstrated by rotating the plant on a horizontal clinostat (Fig. 13) with its axis horizontal, which may reveal the existence of an

Fig. 13. A *Coleus* (Lamiaceae) plant spun on a clinostat develops epinastic leaves. A clinostat is a device used experimentally to negate the gravity vector.

endogenous epinastic component. This procedure, if carried out at the correct angular velocity, exposes the gravity sensors to periods of gravitropic excitation that are too short for the development of a response. Strength of the gravitational component acting on the branches is reduced, favoring the epinastic component. The branches respond by enhanced epinastic (downward) curvature.

The epinastic response is not an inherent feature of the upper sector of plagiotropic branches. In plants that have been grown over extended periods in an inverted orientation (roots up, shoot down), the epinastic response progressively reverses its orientation. Interaction with epinasty is revealed when the plants are inverted vertically (upside down). Epinasty is indicated if downward curvature takes place, because in this orientation it augments the negative gravitropism instead of diminishing it. Alternatively, the signal may originate in the gravity-induced curvature itself. This signal diminishes as curvature is reversed, resulting in straightening. Aerial branches of *Asparagus officinalis* (Liliaceae) are diagravitropic, growing horizontally, perpendicular to the erect, primary axis. Displacing the plant through 90° (primary stem horizontal) results in the branches curving backward toward the base of the plant, with their upper (adaxial) side convex.

Assuming that in such diagravitropic shoots the gravity vector is inherently maximal at right angles to the axis of the branch, displacing it by 90° would minimize it, enabling the epinastic component to exert its effects. Inverting excised rhizomes of *Aegopodium podagraria* (Apiaceae) by rotating them 180° around their axis results in downward curvature, followed by a number of upward/downward oscillations, before resuming diagravitropic orientation. In the inverted rhizome, the epinastic (downward) curvature is initially augmented by the persistent negative gravitropism (downward) until the latter is dissipated and replaced by a fresh negative gravitropic curvature (upward). Plants of *Coleus blumei* (Labiatae) increased the epinastic curvature of their branches and leaves on a horizontal clinostat that was rotating between 4 and 1/3 rpm. This has been attributed to the increase in epinastic curvature under reduced gravitropic conditions. Therefore, the Ax does not appear to be set by gravity alone.

An additional complication with the concept of a dominant GSA is the fact that the spatial orientation of shoots branches of higher order is not commonly consistent with any coupling with the gravity vector. Instead, the orientation is entirely consistent with an endogenous control. They grow out at predetermined Ax, similar to the Ax of the primary branches, and at different, predetermined Az. As a result, they are oriented at a different GSA. These orientations may therefore be autotropic, rather than gravitropic.

The Ax is determined in the early stages of development of the axillary bud. The first buds that become active at the base of the primary shoot of *Sorghum halepense* (Poaceae) are destined to develop into diagravitropic rhizomes. Their Ax becomes progressively more obtuse during development, apparently because the node above them becomes thicker, tilting them outward.

Rhizomes of *Eleocharis palustris* and *Scirpus maritimus* (Cyperaceae), *Aegopodium podagraria* (Apiaceae), *Sparganium racemosum* (Sparganiaceae), and possibly other diagravitropic organs may have structural radial symmetry, but their dorsal (lower) and ventral (upper) sectors exhibit differential physiological responses (physiological dorsiventrality). For instance, adventitious rooting is virtually restricted to the lower sector. Isolated rhizomes of *Eleocharis*, when planted obliquely upward or downward, curve downward or upward, respectively, to resume horizontal (diagravitropic) movement. However, if they are inverted by 180° rotation around their axis, they curve upward as the previously ventral (lower) sector elongates at a higher rate. Minor down/up fluctuations are followed by return to diagravitropic (horizontal) movement. Runners of *Rubus caesius* (Rosaceae) exhibit radial structural symmetry; however, when displaced to a vertical orientation they revert to horizontal movement, but invariably with the morphologically upper side convex, suggesting epinastic dominance.

MODIFICATION BY THE ENVIRONMENT.   Light appears to be a potent factor in modifying the spatial orientation of branch shoots of groundnuts (*Arachis hypogaea*). The *Ax* of first-order branches was extremely acute (approaching vertical) in dark-grown plants, and became progressively more obtuse (approaching horizontal) in plants grown in light with increasing irradiance. A similar progressive change in axial angle took place when the plants were grown in increasingly longer photoperiods. Evidently, the overall number of photons intercepted determines the response. It appears that the intrinsic axial angle of shoot branches may be modified by gravity as well as by light, neither of these acting as vectors. By definition, the response of the branches to light and gravity is not tropic but nastic.

Two-Dimensional Movements

Plants such as *Cedrus libani* (Pinaceae) and *Asparagus plumosus* (Liliaceae) exhibit two-dimensional movements by diagravitropic branches well above ground. These are exceptions. Other plants exhibit two-dimensional movements by prostrate, horizontal branch shoots, as seen, for instance, among *Citrullus* spp., *Cucumis* spp., *Bryonia* spp., *Ecballium elaterium* (Cucurbitaceae), *Tribulus terrestris* (Zygophyllaceae), and *Portulaca oleracea* (Portulacaceae). These are mostly annuals, characterized by a primary shoot with telescoped internodes and a taproot. All their leaves are photosynthetic and similar, and their axillary buds are nondormant. The closed ground cover provided by the foliage of such prostrate branches contributes to reducing the competition by neighbors by inhibiting their germination or sprouting.

Function and Patterns

In many perennial plants, the primary function of shoots and roots that exhibit two-dimensional movements is not nutritional. Diagravitropic shoots and roots produce diagravitropic branches. A variety of plants exhibit two-dimensional movements by producing root branches that move horizontally in different directions. Such roots do not function in absorption as long as they continue moving diagravitropically but produce adventitious buds along their length that may remain dormant. The primary function of such roots is territorial invasion and occupation. Their apex eventually reverts to a positive gravitropic movement and grows vertically down, after which they assume normal absorptive function.

Diagravitropic branches of the shoot have a similar function. In general they are produced in leaf axils at the base of an erect, negatively gravitropic shoot with compressed internodes forming a rosette. The buds from which branches eventually develop are closely clustered to form a hub, which may be situated at or just above the soil surface. In other cases, the hub is located entirely within the soil. In either case, the compressed basal internodes are in intimate contact with the soil and produce numerous adventitious roots. Axillary buds at the hub may grow into erect (orthogravitropic), leafy branches. Such branches also have compressed basal internodes that produce adventitious roots, contributing to the fibrous root system. Thus, the hub represents the epicenter for the meristematic activity involved in production of axillary buds and adventitious roots.

Some axillary buds in the hub metamorphose into branches that grow in different directions at right angles with the gravity vector, exhibiting two-dimensional movements, the result of a diagravitropic response. In some plants belonging to these categories, such horizontal branches develop into runners with very long, slender internodes that remain and creep above and along the soil surface. Their nodes carry vestigial leaves and axillary buds, and usually produce adventitious roots. In other plants, such horizontal branches grow entirely within the soil, below its surface. They exhibit one of two diametrically opposite developmental patterns. In one, they metamorphose into stolons, which are the subterranean counterparts of the aerial runners. Their internodes are long and slender, and their nodes carry vestigial scale leaves and produce adventitious roots. In the alternative developmental pattern, they metamorphose into rhizomes with very short, thick internodes that are loaded with stored food (primarily carbohydrates), and their tightly compressed nodes carry scale leaves with dormant axillary buds. The apical bud of runners, stolons, and rhizomes eventually reverts to a negative gravitropic growth (in response to an environmental signal, or as a result of some endogenous process such as initiation of reproductive development) and grows into an erect, leafy shoot with adventitious roots at its base.

Several plant species produce extensive systems of horizontal roots. These fall into two functional categories. One root system functions as a means for rapid, extensive expansion and occupation of surface areas of surrounding land, in which the roots can then proliferate by means of adventitious buds. Their function is analogous to that of runners and stolons, which do the same by means of specialized, axillary regeneration buds. Plants in the other category employ horizontal roots as platforms for negatively gravitropic movements for the specific purpose of exclusive survival in a specialized, extremely hostile soil environment.

Topographic Movement

The primary role of two-dimensional movement is to invade and occupy adjacent territory by means of regeneration buds from horizontal roots shoots. However, on slopes or where surface topography is otherwise uneven, production by such horizontal organs of viable, well-established new individuals from their regeneration buds may encounter difficulties. The buds may sprout at soil depths that are too great for shoot emergence, or too far above the soil surface for rooting and establishment.

Diagravitropic shoots and roots have evolved a strategy by which they cope with uneven terrain, avoiding these shortcomings. They do so by modifying their diagravitropic response to enable them to move topographically, following the contours of the soil surface below them, or within a narrow range of depth below the surface. Moving topographically maintains the regeneration of buds within an optimal range of distances below or above the soil surface. The underground range represents the physiological depth, which is the depth that is optimal for the establishment of the new plants. These organs are exhibiting gravitopographic movement.

In order to follow the topography of the soil surface, such organs must be able to sense the distance from the soil–air interface and change their direction of growth accordingly. Horizontal growth of diagravitropic branch shoots and roots that move topograhically is controlled primarily by gravity, but the movement of subterranean organs to the physiological depth is probably guided by an ability to sense the depth-dependent gradient of some critical environmental factor such as light, temperature and the amplitude of its diurnal fluctuations, or levels of oxygen and carbon dioxide. These environmental variables may modify the inherent gravity-driven response.

The aerial environment offers few indicators of distance above the soil surface, because the gradients are too small. One notable exception is light. Runners lie prostrate on the soil surface, exposed to the aerial environment and to unilateral light (Fig. 14). The ability of a runner to follow the contours of the soil surface and become appressed to it may be accounted for by the marked difference in irradiance intercepted by the exposed and shaded sectors of its apex. This gradient may induce a negative phototropic response that modifies the inherent diagravitropic response, leading to curvature away from the light source and directing growth

Fig. 14. Strawberry plants have a runner—a specialized stem that grows horizontally away from the parent plant and produces a new plant. This is one example of vegetative propagation and colonization of territory by plants.

toward the soil surface. Unable to overcome the mechanical resistance of the soil, the runner assumes a prostrate growth habit.

The shoot of ivy, *Hedera helix* (Araliaceae) exhibits interaction between its inherent negative gravitropic response and a negative phototropic response to the blue region of the visible spectrum. It climbs up opaque objects such as walls, rocks, and tree trunks by becoming appressed against them. It then adheres to them by means of a bristle of stiff adventitious roots that develop only on the shaded side. In the absence of light, the negative gravitropic response of the shoot predominates, making it grow upward without being appressed. Runners of *Alternanthera ficoidea* (Amaranthaceae), *Commelina cayennensis* (Commelinaceae), *Portulaca oleracea* (Portulacaceae), *Citrullus* spp. (Cucurbitaceae), and *Echinochloa colona* (Poaceae) maintain the prostrate habit of the shoot only in light, becoming negatively gravitropic in darkness.

CHAPTER FOUR

The Underground World

A IR IS TRANSPARENT, but soil is opaque, and this is one reason why we are only vaguely aware of just how much goes on within the soil. That many animals use the soil as their home is familiar, as is the obvious fact that soil is the natural habitat for most plants' roots. Aerial roots of epiphytes and the pneumatophores of mangrove plants are a notable exception. However, it is less widely known that many plants also use the soil as the preferred shelter for some of their most vital developmental assets—the tender meristems that produce all their aerial organs, including their photosynthetic foliage, and eventually their reproductive organs (flowers and inflorescences) and offspring (seeds). These plants are called the geophytes.

With few exceptions, the seeds of all plants germinate at the soil surface or just under it, thereby exposing their shoot apices to a harsh environment. Geophytes have evolved strategies for moving the shoot apices of their seedlings and their meristems into the much more hospitable environment inside the soil, thereby enhancing protection throughout their subsequent vegetative activity. The shoot apices usually emerge above the soil surface only in conjunction with onset of their reproductive development and transformation into flowers and inflorescences. Some of these plants have evolved mechanisms for moving these subterranean assets over considerable distances and in specific directions through the soil for a variety of purposes. However, the structure of soil presents complex mechanical resistance to the progress of delicate plant organs (Fig. 15).

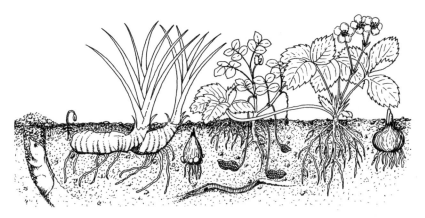

Fig. 15. The underground world is full of plant organs, roots, stems, and other parts moving to optimize their physiological depth.

Physiological Depth

Plant parts that inhabit the underground world face an entirely different environmental complex than their aerial counterparts. Their ability to grow underground requires that they have an altimeter or depth gauge, but very little information is available on the components of such sensors. The underground world is characterized by steep gradients in biologically important environmental variables with increasing depth within the soil. Topographic movement through the underground world may be accounted for by a sensory mechanism capable of discriminating environmental variables whose steep gradients within the soil are reliable indicators of depth. Such a mechanism, located in the shoot apex of the subterranean shoot, may guide the movement of the shoot within its physiological depth.

The subterranean environment is conceived as a dark place, but light does penetrate below the soil surface. Cracks result from contraction of the soil as it dries, and animal activity creates additional avenues for penetration by light. The soil matrix consists of mostly macroscopic crystalline particles separated by interstices of various dimensions, so the light that does penetrate below the surface may travel considerable distances (as much as 10 cm) within the soil by multiple reflections from crystalline soil particles as well as by other tortuous pathways. The irradiance is strongly attenuated as it travels into the deeper soil layers, but many sensory systems that are involved in photomorphogenesis exhibit extreme sensitivity to light. They may be initiated by light of extremely low irradiance or of very short durations. A single exposure of the rhizome of

Aegopodium podagraria (Apiaceae) to 30 seconds of red light causes downward curvature, followed by upward curvature, before reverting to diagravitropic movement. Rhizomes of Solomon's seal *Polygonatum multiflorum* (Liliaceae) curve downward in the dark and upward in light. Stolons of *Ajuga reptans* (Lamiaceae), *Ranunculus repens* (Ranunculaceae), and *Hieracium pilosella* (Asteraceae) exhibit similar responses to light (downward curvature) and dark (vertical upward). Downward curvature in response to light increases with irradiance; the response is not ascribed to negative phototropism but rather to a tonic (nondirectional) enhancement by light of the inherent epinastic response. Clearly, light may be included as a possible (tonic) parameter by which subterranean organs determine the physiological depth.

In addition to light, the partial pressure of oxygen is also strongly attenuated with increasing depth in the soil, while the partial pressure of carbon dioxide increases. Exposure of the rhizome of *Aegopodium podagraria* (Apiaceae) to an atmosphere of 5 percent carbon dioxide causes it to curve upward.

The level of temperature and its diurnal amplitude below, above, and especially at the soil surface differ significantly and are strongly dependent on the presence or absence of direct solar radiation as well as on the topography of the soil surface. Soil surface temperature under solar radiation is considerably higher than in the air above it, where sensible heat is carried away by convection and wind. It is also much higher than in the soils underneath as a result of the high heat capacity of soil, especially when wet. The opposite situation occurs at night, where the surface cools off rapidly by radiation to the cold sky. As a result, the amplitude of the day/night temperature fluctuations is most extreme at the surface and declines rapidly with depth under the surface. Response to thermoperiodicity is a well-known phenomenon in a variety of morphogenetic processes, such as seed germination, bud dormancy, and induction of flowering.

All these environmental parameters, as well as others that are less critical, exhibit a pronounced gradient with increasing depth within the soil. As a result of these gradients, the complex of environmental variables may change significantly within a very narrow range of depth within the soil. This has led to the concept of a physiological depth, where the complex of environmental variables is optimal for the well-being of subterranean regeneration buds and is species specific.

Coping with Mechanical Resistance

Apices of the root and shoot lead the advance through the soil against considerable mechanical and frictional resistance by the soil matrix, yet they contain the apical meristems—the youngest, most delicate tissues—and, in the case of the shoot, the leaf primordia and youngest leaves. The motor tissue for this movement is the subapical elongation zone of the root or stem, which propels the apices of these organs through the soil against the mechanical resistance by the soil's solid particles. Roots and subterranean shoots have evolved mechanisms for navigating through the soil environment. The specialized structure and function of their apices provide the necessary protection for these essential, delicate tissues.

Roots

A highly coordinated apparatus facilitates penetration of the root tip through the soil. Structural and functional features of the root tip facilitate its movement against the frictional resistance of the soil. The root tip is characteristically shaped as a blunt-headed bullet, which enables it to push aside soil particles and thus insert itself between them. The apical meristem of the root, situated at its very tip, adds new cells by periclinal divisions toward the base of the apex. Thus, the tip is inexorably propelled forward by its own product while at the same time it expands its girth from the tip toward the base. These two forces combine to act as a wedge that moves aside, slowly but relentlessly, the neighboring soil particles by exerting linear and radial pressure.

In addition, the root tip has evolved a strategy to cope with the hazards presented by friction caused by movement between the tightly packed soil particles. A special protective cup-shaped organ—the root cap (calyptra)—fits over the hemispherical root tip, the meristem, like a cup for an egg. The root cap keeps on adding new cells acropetally by means of periclinal divisions of its meristem (calyptrogen), which adjoins the apical meristem of the root at the base of the cup.

Like all cells, those of the root cap adhere to each other by means of the pectin in their middle lamella. Pectin is an amorphous, strongly adhesive substance formed from an indeterminate number of molecules of pectic acid (a linear polymer of uronic acids, primarily galacturonic acid) bound to each other by the divalent cations calcium (Ca^{2+}) and magnesium (Mg^{2+}). Formation of the salt transforms the highly hydrophilic, gelatinous pectic acid into a very strong adhesive. Pectin, produced by numerous Golgi apparatus, also gradually fills up the lumen of all cells of the calyptra. As these cells mature, the pectin is hydrolyzed into a watery slime, no longer capable of holding the cells together. While this goes on, these cells are expelled toward the perimeter of the cap by the new cells that continue to be added behind them. They are eventually shed around the exterior of the root cap. As they are shed, cap cells become extremely fragile and readily break open, releasing their stored pectic acid. This concerted activity provides the root cap with a highly effective lubricant that facilitates its penetration between the soil particles with minimal resistance.

Subterranean Shoots

A specialized structure facilitates movement of the subterranean shoot apex through the soil: the apical bud (plumule) of subterranean shoots is also bullet shaped. The leaf primordia are consecutively formed close together, tightly appressed over the apical meristem. The older, more mature primordia clasp all those that are younger and less mature, forming a compact, closely fitting conical shield over them and over the apical meristem at their center. Eventually the oldest, most mature internode elongates, propelling the apical bud forward and leaving behind the oldest, most mature leaf primordium at its base. Leaf primordia are produced and develop basipetally. As they expand and thicken, the diameter of the apical bud increases toward its base, thus acting as a wedge that forces aside the soil particles, similar in function but differing in operation from the root apex. This strategy enables the apical bud as well as its expanding leaves to minimize damage by any obstacles they might encounter as they are propelled forward.

Leaves produced on nodes of subterranean shoots do not expand and do not develop any photosynthetic function. Instead, they mature very early in their existence into vestigial leaves or tough, resistant bracts that offer protection to the buds in their axils as the elongating shoot moves through the soil. Prostrate shoots produce leaves with similar characteristics that cope with the resistance of sliding along the soil surface and its superficial detritus.

Strategies for Moving Seedlings through the Underground World

Most seedlings start their existence entirely within the soil. Growth of their roots propels the apical meristems at their tips ever deeper into the soil. The structure of the root apex is compatible with penetration into the soil. This is not the case with the seedling's growing shoot. The young shoot of the dicotyledonous seed is faced with greater difficulties. It is headed by the delicate apical bud and (in epigeal seedlings) by two large, unwieldy cotyledons. Its unfolding cotyledons and/or expanding leaf primordia not only increase the resistance to penetration, but also are also extremely fragile. Seedling shoots have evolved strategies by means of which their cotyledons and/or young leaves minimize the mechanical resistance of the soil that they encounter as they move upward on their way to emergence. They do this by means of structural modifications that enable them to bypass contact with the soil matrix. Furthermore, the environment within the soil is progressively more precarious and hazardous for the tender meristems the closer they are to the soil surface. This is especially the case where the soil is exposed to direct sunlight during the day and the cold open sky at night, making temperature and its diurnal fluctuations more extreme and depletion of soil moisture more rapid.

Plants have responded to these hazards by evolving developmental strategies for moving their core meristems into safer, more stable environments deeper within the soil. All seedlings that germinate inside the soil move though the soil, but movement of dicot seedlings differs fundamentally from that of monocot seedlings.

Emergence of Dicotyledonous Seedlings

Dicot seedlings move their primary root apex vertically down and the apical bud of their primary shoot vertically up through the soil, but they maintain their base in its original location and depth. The most prevalent modification in shoot organs of seedlings of dicots is the hook (Fig. 16). In darkness, seedlings are etiolated. A well-defined segment of the distal part of the spindly, elongated hypocotyl/epicotyl subtending the hook consists of the actively elongating tissues and is positively gravitropic. It is curved down at an acute angle so that the cotyledons and/or the apical bud and the unexpanded young leaves it contains are pointed straight down, parallel to the subtending part of the stem (hypocotyl or epicotyl). Thus, the hook acts as a wedge propelled forward through the soil by elongation of the zone that subtends it. The apical bud and its foliar appendages are dragged base first, along the pathway cleared by the advancing hook.

The hook itself is the specialized distal part of the zone of elongation. It is formed by highly differential elongation, inhibited strongly in its lower (concave) sector. This inhibition is temporary; as the tissues mature, elongation of the lower (concave) sector reverts, overtaking that of the upper (convex) sector, which results in straightening the basal (more mature) part of the hook. In the meantime, the youngest parts of the zone of elongation, beyond the hook, now adopt the differential growth and join the hook. The hook remains operational only as long as the shoot advances through the soil or in darkness.

When greening—that is, conversion to de-etiolated form—is initiated by exposure to red wavelengths of light, stem elongation is repressed, the

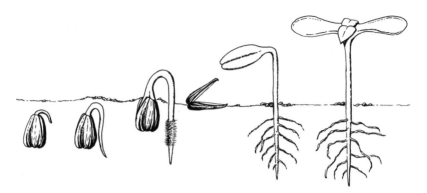

Fig. 16. A germinating dicotyledonous seedling emerges.

hook straightens rapidly, and the cotyledons and/or expanding leaves assume their normal orientation, face upward. This action of light takes place by phototransformation of the inactive form of phytochrome, P_r, to its active form P_{fr}. The hook serves to facilitate the movement of the apical bud of dicot seedlings through the soil, the shoot apex eventually ends up above the soil surface, and the liberated apical bud now can unfold its leaves to the light and air.

Emergence of Monocotyledonous Seedlings

In monocots, the seedlings use the cotyledon or (in the grass family, Poaceae) the primary embryonic leaf as a means for enabling their delicate, elongated foliage leaves to emerge from the soil without actually coming into contact with it. In the grass family (such as oats, *Avena* spp.; barley, *Hordeum* spp.; and maize, *Zea* spp.), the entire single cotyledon (scutellum) remains embedded within the endosperm, where its role is the mobilization of the food stored in the endosperm and its uptake for use by the seedling. Therefore, by default, the primary embryonic leaf is relegated to the role as tool of preference for facilitating emergence of the young leaves from the soil.

This embryonic primary leaf consists entirely of a tubular sheath devoid of a lamina, with a bullet-shaped tip, that encloses the plumule situated at its base. Its unique structure and function have been recognized by a special name: coleoptile. Its subtending (first) internode has been named mesocotyl in recognition of its own special status and function. Germination of the grass seed starts with emergence of its radicle from its sac-like sheath (coleorhiza), which is followed by emergence of the coleoptile and its subtending mesocotyl. Both are negatively gravitropic; the coleoptile tip moves upward, toward the soil surface, and the bullet-shaped tip and smooth exterior of the coleoptile contribute to its ability to overcome the mechanical resistance of the soil.

At the same time, elongation of the negatively gravitropic mesocotyl (which is suppressed in cultivars of wheat, *Triticum* spp., and barley, *Hordeum* spp.) propels the entire elongating coleoptile upward. The coleoptile lacks meristematic activity. Its growth is entirely dependent on elongation of existing cells, which places an upper limit on its final length. The mesocotyl, on the other hand, contains an intercalary meristem, albeit of limited life span, that enables it to continue growing long after the coleoptile at its top has exhausted its own growth potential. Elongation of the mesocotyl can be quite considerable in primitive (Native American) maize (up to 30 cm). The mesocotyl is thus able to continue propelling the coleoptile toward the soil surface.

In addition to its role in facilitating emergence of the leaves from the soil, the coleoptile provides the grass seedling with a means for controlling the optimal depth within the soil depth at which the base of the seedling shoot comes to rest and develops into the mature shoot. All growth activity of the coleoptile and mesocotyl comes to an abrupt end as soon as the tip of the coleoptile emerges from the soil and is exposed to light. When this happens, the leaf primordia in the apical bud, whose growth had until then been suppressed, start elongating through the smooth-walled coleoptile and eventually emerge from its tip without having touched the soil. At the same time, adventitious roots start to develop at the base of the coleoptile (coleoptilar node).

In most monocots other than grasses, the seed contains a single cylindrical cotyledon consisting of two parts. The distal part is solid and remains embedded within the endosperm of the seed (at least until the seedling becomes established). It transforms into a tissue that functions in the mobilization and absorption of the stored food. The proximal portion of the cotyledon consists of a hollow, smooth-walled tube, with the plumule at its base and the radicle at its tip. This part is positively gravitropic and moves straight down. The seed germinates by emergence of the embryonic radicle, which is positively gravitropic, its apex propelled downward by elongation of the cylindrical cotyledon, thrusting aside the surrounding soil matrix.

Development of the radicle is arrested as long as the cotyledon propels it downward through the soil. The compact structure of the inactive radicle enables it to function as the tip of the wedge that facilitates the movement of the elongating cotyledon through the soil. When the radicle starts growing, the primordial leaves of the apical bud start elongating upward within the hollow cotyledon and eventually emerge through it above the soil surface. The emerging cotyledon exhibits gravitropic responses, which differ in species with hypogeal and epigeal seedling emergence. In the hypogeal seedling, the entire cotyledon is positively gravitropic and grows vertically down, without appearing above the soil surface.

The downward movement of the base of the cotyledon moves the apical bud to a suitable depth underground, where the adventitious root system will also be established. Palms (represented by the date palm *Phoenix dactylifera*, Arecaceae) fall into this category. In epigeal seedlings, the cotyledon is initially negatively gravitropic and grows vertically upward. Soon afterward, it develops a sharp hook at the transition between the distal, solid part and the tubular, proximal part. At first, both parts elongate, propelling the hook upward until the entire cotyledon is above the soil surface, anchored to the soil by the radicle. Continued intercalary elongation of the proximal part acts as the motor that propels its tip, headed by the radicle, vertically down into the soil. Within the soil, the sharp hook facilitates movement against the frictional resistance of the soil. Soon the distal part stops growing, while the proximal part continues to do so. This differential elongation causes the distal part to extract from the seed coat as well as from the soil.

Eventually the tip of the cotyledon reaches its proscribed depth; the radicle and the young leaves start growing, and adventitious roots start developing. The first young leaf elongates through the tubular cotyledon, which it breaches above the soil surface, thus avoiding any contact with the soil and its friction. The young leaf is rolled longitudinally and performs the same service for its younger followers. Members of the Liliaceae family such as the domestic onion *(Allium cepa)* and garlic *(A. sativum)* fall into this category. As in the grasses, this elaborate mechanism determines the soil depth at which the base of the seedling shoot comes to rest and develops into the mature shoot (including a bulb, or corm). Seedlings of a number of species of *Allium* establish their bases at a depth of ~50 mm. Those of *A. schubertii* may go as deep as 170–200 mm.

The tulip (*Tulipa* spp., Liliaceae) exhibits epigeal germination similar to the onion and garlic, but it introduces a novel improvement. The proximal, tubular part of the cotyledon inserts its tip, containing the apical bud and embryonic radicle, straight down into the soil. At this point, it develops an additional burrowing organ. This special organ is also a hollow tube, with the apical bud and embryonic radicle at its tip, but it consists of tissues of the base of the cotyledon as well as vascular connections to the base of the cotyledon. This "relay" organ is in series with the cotyledon, and like the cotyledon, it exhibits a positive gravitropic response, growing straight down into the soil, increasing the depth at which the new bulb becomes established.

Movements of the monocot seedling serve an additional function. They end with the apical bud sheltered within the soil and protected within the sheath of its surrounding leaves. The elaborate interplay between elongation of the coleoptile of the grass seedling with its light-sensitive tip and the mesocotyl sets an upper limit to the soil depth at which the bud at the base of the seedling shoot comes to rest. This allows a hub to form, from which the foliage leaves of the vegetative shoot emerge and an adventitious root system develops. This depth corresponds to the maximal capacity for elongation of the coleoptile. Monocots other than grasses exhibit either hypogeal or epigeal seedling emergence. Their seedlings establish their hub at a depth determined by the capacity for elongation of the positively gravitropic part of their cotyledon. It is from this depth that the leaves sprout and emerge from the ground as part of an erect, leafy shoot.

Shoots of most monocots are characterized by a number of closely compressed internodes at their base. The associated nodes are capable of producing a number of adventitious roots. The plumule remains underground, protected by the concentric sheath formed by the surrounding leaf bases, until the upper (more recently formed) internodes of the shoot elongate (usually at flowering). The subterranean nodes produce axillary buds that remain protected by a layer of soil. Eventually they develop and emerge as aerial branches that are clustered close together. In grasses, the erect basal branches are named tillers. In the palm *Phoenix* and the banana *Musa* spp. (Musaceae), they are named offshoots. Their basal internodes are similarly compressed and are fed by their own set of adventitious roots.

The ability of many monocots to move the apical bud of their seedlings underground is the primary factor that enables them actively to affect and even determine the maximal depth within the soil at which the entire vegetative meristematic activity of the whole plant (the hub) becomes established. This is how grasses and other moncots occupy densely the uppermost layers of the soil up to a specific depth, depending on the species.

Interment in Dicots

Some dicot seedlings protect their shoot apices underground as a strategy for sheltering their apical buds within the soil. Most commonly, the petioles of the two cotyledons fuse into a tubular sheath that surrounds the apical bud. The petiolar sheath formed by their fusion is positively gravi-tropic, elongating downward by intercalary growth, thus propelling its base, with its enclosed apical bud, into the soil. Growth of the embryonic radicle is inhibited while the petiolar sheath is elongating; its tip acts as a wedge that facilitates the advance of the petiolar tube through the soil. Young leaves grow up through the tube into the free atmosphere without encountering the soil at all. *Eranthis hyemalis* (Ranunculaceae), and *Leontice leontopetalum* and *Bongardia chrysogonum* (Berberidaceae) represent this category. The shoot apex of the latter two eventually attain depths of 300–350 and ~250 mm, respectively, after which they transform the base of their hypocotyl into a corm. In all these categories, the cotyledon, despite being part of the shoot, exhibits a positive gravitropic response and emulates the gravitropic response of the root at its base.

Some species have evolved strategies for protecting their offspring in the safety of the underground before they are transformed into achenes (single-seeded dry fruit of the Asteraceae). For example, *Gymnarrhena micrantha* (Asteraceae) is a dwarf desert annual that practices amphi-carpy: it produces two distinct types of inflorescence, differing also in the morphology and function of their fruit. Exposed inflorescences are lo-cated at the soil–air interface and carry numerous small achenes equipped with a pappus that aids their dispersal by wind. Subterranean inflores-cences are few in number and carry one or two large achenes (no pap-pus), each at a depth of ~1–2 cm underground.

The exposed achenes are dispersed far by means of their pappus. They germinate on top of the soil, and their radicle develops into a taproot. In the course of its development, the taproot starts to contract acropetally, pulling behind it the apical bud into the soil. As this process goes on, axil-lary buds grow upward through the tunnel left by the receding apical bud; on reaching the surface, they develop into the exposed inflorescence branches. The subterranean apical bud develops into an inflorescence that produces a small number of large achenes. These remain protected inside the soil as well as by the dead, tough tissues of their parent. They germi-nate in situ, still enclosed within the dead tissues of their parent and in the precise location where their parent successfully completed its life cycle, a distinct advantage in deserts. A less elaborate amphicarpy is also exhibited by *Vicia sativa* and *Pisum fulvum* (Fabaceae), *Catananche lutea* (Astera-ceae), *Linaria* spp. (Scrophulariaceae), and *Emex spinosa* (Polygonaceae).

The groundnut *Arachis hypogaea* (Fabaceae) starts and carries out the entire development of its pods and seeds only after inserting them forcibly to a depth of several centimeters into the soil (Fig. 17). Soon after fertilization, the penultimate floral internode subtending the ovary—the gynophore—starts elongating vigorously, curving straight down (positive gravitropism) and actively inserting the immature ovary at its tip into the soil. Elongation of the gynophore apparently suppresses any growth of the tiny ovary and any development of its enclosed embryo. The small size of the ovary enhances its movement and insertion into the soil.

Development of the ovary into the familiar pod and of its embryo into the familiar seed of the groundnut start as soon as they have reached the predetermined depth of penetration into the soil by the gynophore. During the rapid and vigorous developmental growth, the pod changes its orientation to diagravitropic. Light appears to be a potent factor in modifying the spatial orientation of branch shoots of the groundnut *A. hypogaea*. The axial branching angle (*Ax*) of first-order branches is extremely acute (approaching vertical) in dark-grown plants, and the angle becomes

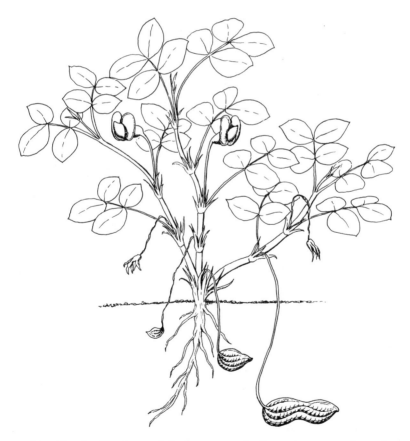

Fig. 17. After fertilization and the interment of a flower by a groundnut plant, the developing groundnut matures underground.

progressively more obtuse (approaching horizontal) in plants grown in light with increasing irradiance. A similar progressive change in axial angle takes place when the plants are grown in increasingly longer photoperiods; evidently, the overall number of photons intercepted determines the response. *Factorovskya aschersoniana* (Fabaceae) exhibits similar behavior. This adaptation not only protects the seeds, but also ensures that the seedlings will benefit from the same environment in which their parent successfully completed its life cycle.

Trifolium subterraneum and *T. israeliticum* (Fabaceae) bury entire inflorescences in the soil. After fertilization, the inflorescence consists of a small number of developing pods, totally enclosed by concentric circles of sterile flowers. The calyx of the sterile flowers develops into long, stiff, sharp-tipped bracts equipped with barbs along their edges. These are curved over the pods at the center of the inflorescence. The entire inflorescence is thus molded into a compact, globular, spiny structure; as the inflorescence matures, its stalk elongates while reversing its gravitropic movement to positive, and it drives the entire inflorescence into the soil, with the inflorescence acting as a drill head.

Some dicot plants have developed analogous strategies for protecting their developing fruits and seeds in relative safety underground. The flower of *Cyclamen persica* (Primulaceae) is borne on a long, flexible stalk that exhibits positive phototropism. Soon after fertilization, the stalk reverses its phototropic response and curves away from light by progressive acropetal growth, until the developing fruit at its head finds a crevice in the soil or among rocks and is thrust inside. The flower stalk of *Cymbalaria muralis* (Scrophulariaceae) behaves similarly, reversing its phototropic response soon after fertilization. The floating water hyacinth *Eichhornia crassipes* (Pontederiaceae) exhibits a similar response: soon after fertilization, the erect stalk of the inflorescence reverses the direction of its movement to positively gravitropic and grows downward, submerging the entire inflorescence, where the fruit and seeds develop and mature.

Problems of Living Underground

For all it attractions, life underground does create its own class of problems. Migration is one example. Plants tend to migrate to habitats that provide them with an optimal environment. Above ground, migration is primarily by means of seed dispersal carried out by wind, water, and a wide variety of animals. To some extent, migration occurs by invasion, occupation, and colonization, using specialized horizontal (probably diagravitropic) lateral branches of shoots (runners, stolons) and roots. Underground migration requires special adaptations.

Geophytes make the underground world their habitat of choice. However, even for geophytes life underground presents a variety of problems. The geophyte shoot starts migrating underground as soon as it germinates, driving its apical meristems through the soil. The apical bud and its shoot meristem keep on moving underground toward or within the range of its physiological depth throughout vegetative activity. They eventually transform to negative gravitropism to produce erect foliage that emerges above ground. The apical meristem itself emerges only after it had initiated reproductive development. These two events appear to be correlated.

Mangroves

Extensive areas along the coastlines of tropical and subtropical regions have become mangles, highly specialized habitats and ecosystems created in river estuaries where fresh water, laden with silt and partially decomposed organic matter, flows into shallow, saline coastal waters. Large stretches of these areas are transformed into brackish swamps with totally anaerobic soil that cannot support normal plant communities. In addition, most of these habitats are intertidal, exposed daily to inundation by seawater and high fluctuations in salinity. This extremely inhospitable habitat has become the sole refuge for a highly specialized plant community of trees and shrubs: the mangroves. There are relatively few plant species in this community, all of whom have evolved the same or similar adaptations for survival, despite their belonging to diverse taxonomic groups. For example, red mangrove (mangle rojo) *Rhizophora mangle*, *R. mucronata*, and spotted mangrove *R. stylosa* are Rhizophoraceae; gray mangrove *Avicennia marina* and black mangrove (mangle negro) *A. germinans* are Avicenniaceae; and mangrove apple *Sonneratia alba* and mangrove crabapple *S. caseolaris* are Lythraceae.

All these mangrove plants fail to develop a primary root. Instead, the seedling shoot produces a vast, widespread system of adventitious roots, all of which move horizontally, just under the surface of the marshy soil, and act as the sole mechanical support for the entire tree/shrub, which may be vast. Tree species may reach heights exceeding 10 m and crown diameters exceeding 20 m. However, the horizontal roots are able to function in the marshy, anaerobic, and saline substrate by producing pneumatophores (breathing roots), which are highly specialized, first-order branch roots that are negatively

(continued)

gravitropic and move vertically upward, above the surface of the swamp for considerable heights (commonly <50 cm, and up to 300 cm). As many as 10,000 pneumatophores have been recorded for a single individual of *Avicennia nitida*. Densities exceeding 500 pneumatophores per square meter have been recorded for *A. marina*.

The internal structure of pneumatophores consists primarily of aerenchyma—a parenchymatous tissue characterized by a widespread system of interconnected intercellular spaces. The pneumatophore is externally protected and isolated by an impermeable layer of cork cells, but it maintains direct contact with the external air via lenticels, specialized openings in the corky layer. These provide the entire submerged root system with the oxygen that is required for survival and function.

In *Bruguiera cylindrica, B. gymnorhiza,* yellow mangrove *Ceriops tagal, and C. decandra* (Rhizophoraceae), the horizontal root abruptly moves upward into the free air, then just as abruptly changes direction and moves back into the ground, forming knee roots that perform as pneumatophores (complete with aerenchyma and lenticels). *Rhizophora* and other mangrove species also produce plagiotropic adventitious prop (or stilt) roots from the aerial parts of their trunk. These arch down into the marsh soil, where they become firmly anchored at some distance from the tree while growing in thickness; they form flying buttresses that provide mechanical support in the highly unstable, marshy soil. Prop roots are also equipped with lenticels and aerenchyma.

Geophytes with bulbs or corms inevitably move outside the range of their optimal physiological depth—that is, the depth in the soil at which physiologically the organ functions best. Shoots of these storage organs are erect. Their regeneration buds develop in axils of scale leaves situated close to the apical bud, above the storage organs. As a result, each generation finds itself at shallower depths than its predecessor, moving progressively out of the physiological depth. The physiological depth itself may also be displaced haphazardly when external agents, such as erosion by wind or by surface runoff, modify the topsoil. This problem is aggravated in moving sands and dunes. Changes in density of vegetation may displace the physiological depth by changing the temperature regime and/or humidity inside the soil. Flooding and drying may change compaction of the soil, and so on.

Certain geophytes produce specialized, horizontal branch shoots or roots adapted to move diagravitropically underground (roots, rhizomes, and stolons) or at the soil/air interface (runners) while producing regeneration buds (axillary or adventitious) along their entire length. Their prime function is territorial invasion and occupation, by moving away from the parent plant, using these buds to establish erect leafy shoots at increasing distances from the original parent plant. Their success depends on their sprouting and producing adventitious roots within a permissible distance from the soil surface. Otherwise, they jeopardize the probability of successful establishment. This distance represents the physiological depth for these plants. These intrinsically horizontal organs face problems where the soil surface departs significantly from the horizontal.

Geophytes cope with these problems by their ability to sense the deterioration of their present environment and relocate themselves, or at least their more vital parts, toward a more favorable environment. In order to relocate, geophytes have evolved a variety of motors specializing in moving bulky and unwieldy objects through the underground world with its inherent mechanical resistance.

Navigation

Moving an object requires force, as well as an anchor on which this force is applied. Apices of growing shoots and roots are structurally (and functionally) adapted to penetrate the soil matrix. The soil offers considerable mechanical resistance to movement. Therefore, moving the apex of any root or subterranean shoot through the soil matrix requires use of considerable force as well as a firm anchor. The force for propelling the apex of the shoots or roots through the soil matrix is generated by axial growth of tissues that subtend them, acting as the motor. The more mature tissues behind the motor provide the anchor.

For roots, the soil is their natural environment, and they are inherently equipped to move through the soil, using the root cap (calyptra) as a shield and lubricant. They are also typically more slender and flexible than the subterranean shoots, and therefore are easier to maneuver around small obstacles.

The force required to move shoots depends on the resistance, which is a function of the mechanical properties and moisture content of the soil on the one hand, and on the size and mass of the moving shoot on the other hand. Bulbs and corms increase in bulk as they develop, and their increasing bulk results in progressive increase in the mechanical and frictional resistance to their movement within the soil. Therefore, their propulsion forward by means of a subtending motor is most effective during the formative stages of their development as propagules (bulbils or cormels).

Stolons are characteristically slender, and their apices face essentially the same resistance as roots in moving through the soil until they transform into tubers and stop moving. Apices of rhizomes are considerably thicker, which presents greater resistance by the soil matrix to penetration, but their subtending motors are equally thick, which increases the forward pressure. On the other hand, bulbs and corms are too bulky to move by such forward propulsion, so they require other means for relocation.

Stolons, rhizomes, and runners also differ from bulbs and corms in their gravity-sensing apparatus. Bulbs and corms are typically erect, and their sensory apparatus is at equilibrium when their axis is aligned along the gravity vector. In contrast, runners, stolons, and rhizomes are diagravitropic, and they inevitably move perpendicular to the gravity vector. Their gravity-guided sensory apparatus perceives, transduces, and transmits the signals to the motor where they control differential growth of the different sectors. However, if their sensory apparatus experiences the direction perpendicular to the gravity vector as an equilibrium state, the same signal is transmitted to all sectors of the motor, resulting in linear movement. The direction of growth remains unchanged as long as the transduced signal is at equilibrium, evenly distributed over the periphery of the target tissues. Changes in direction take place when the transduced signal is differentially distributed on different sectors of these tissues, which results in differential forces and curvature in a direction that can eventually restore equilibrium in the sensory apparatus.

Invasion, Occupation, and Colonization

Runners and stolons are characterized by long, slender internodes, dormant regeneration buds in the axils of vestigial leaves, and an active apical bud that moves diagravitropically and plays a major role in providing a highly efficient means for the plant to invade extensive areas in its habitat. Their exaggerated elongation, rapid growth, and vestigial leaves enable them to move over great distances along or at some distance below the soil surface.

The erect, leafy shoots that develop from their axillary buds root adventitiously, transforming them into autonomous plants that remain interconnected and benefit from mutual support. Thus, invasion graduates into occupation. Furthermore, internodes of runners and stolons are characteristically thin, fragile, and short-lived. They eventually senesce and disintegrate. The autonomous plants achieve independence, and occupation graduates into colonization.

Eventually, the distal, actively growing part changes its diagravitropic response to either a negatively gravitropic shoot or a positively gravitropic root by curvature of its subapical zone of active elongation. Developing rhizomes of *Sorghum halepense* (Poaceae) maintain their horizontal orientation (diagravitropic) for a considerable period in their ontogeny. Eventually, the distal younger, growing parts exhibit positive gravitropic competence by upward curvature.

Runners, represented by *Fragaria* spp. and *Potentilla reptans* (Rosaceae), *Ranunculus repens* (Ranunculaceae), *Paspalum dilatatum* (Poaceae), *Oxalis corniculata* (Oxalidaceae), and *Trifolium repens* (Fabaceae), eventually convert their apical bud to orthogravitropic movement in response to some endogenous or environmental signal. The leafy, erect shoot roots adventitiously and produces a new generation of runners from its axillary buds. These structural and functional characteristics are ideal for invasion of a wide surface area by moving rapidly across or just under the soil surface to establish new, widely spaced, autonomous individuals.

Most stoloniferous plant species produce mixed subterranean diagravitropic shoots with sections with long, thin internodes that are devoid of regeneration buds, typical of stolons and dedicated to long-distance movement, terminated by sections that carry dormant regeneration buds which are rich in stored food and are dedicated to sustained regeneration. The food may be stored in a tuber, consisting of several thick, short internodes with vestigial scale leaves, or in a bulb, consisting of a rosette of thick, fleshy scale leaves. Internodes of the tuber are progressively more compressed, terminated by the apical bud. The apical bud and its most proximal axillary buds are usually those from which new erect leafy shoots eventually develop. The apical bud of a bulb usually ends up as a flower or inflorescence. The new generation of bulbils is produced from axillary buds. These developmental patterns are quite flexible.

Cyperus esculentus and *C. rotundus* (Cyperaceae) produce typical stolons in the axils of the rosette of leaves on the erect shoots. The apices of these stolons may change their diagravitropic movement to an orthotropic one and give rise to new erect, leafy shoots, or they may metamorphose into tubers. The tuber of *C. rotundus* differs from that of *C. esculentus* in that one or more of its dormant buds may grow into stolons that repeat the pattern, forming chains of tubers and leaving the remaining buds dormant for extended periods. The stolon of *C. rotundus* also differs from that of *C. esculentus* by its capacity to form lateral buds in the axils of its vestigial scale-leaves, from which branch stolons develop.

A tuberization pattern similar to that of *C. esculentus* is exhibited by a number of hydrophytes. *Nelumbo nucifera* (Nymphaeaceae) forms diagravitropic stolons in which, at the end of the growing season, the terminal one or two internodes form a tuber. In the following spring, the terminal bud resumes growth as an erect shoot. *Scirpus grossus* (Cyperaceae) and a number of *Sagittaria* (Alismataceae) species produce numerous typical stolons. During the growing season, the apical bud may change its diagravitropic growth into orthotropic growth and give rise to new, erect leafy shoots. Toward the end of the growing season, stolon apices become positively plagiotropic, moving deeper into the soil where they metamorphose into tubers. In the submerged aquatic genus *Potamogeton* (Potamogetonaceae), lateral buds at the distal ends of the diagravitropic shoot systems are transformed into tubers. The apical bud of the tuber sprouts in the following season, forming a leafy shoot.

Allium ampeloprasum (Liliaceae) forms a perennial bulb, from which numerous stolons are produced in the axils of its fleshy scale leaves. These are terminated by bulbils that may remain dormant for a number of years. *Oxalis latifolia* (Oxalidaceae) also forms bulbs, whose apical buds produce foliage leaves and terminate by an inflorescence. Axillary buds in this bulb grow into typical stolons, whose apices form primary bulbils. These may, in turn, form branch stolons with secondary bulbils at their tips.

Stachys palustris and *Mentha arvensis* (Lamiaceae) represent a different category of subterranean diagravitropic shoots. These are formed from axillary buds at the base of the erect shoot. The proximal regions of these shoots are stoloniferous and ephemeral, while their distal regions are rhizomatous and perennate.

Horizontal roots also provide highly efficient means for invasion and occupation of extensive territory. They produce numerous adventitious regeneration buds, either at random locations or in developmentally determined locations. As long as they maintain horizontal movement, they do not function in absorption but rather as producers of adventitious buds. *Ailanthus altissima* (Simaroubaceae), *Prosopis farcta* (Fabaceae), *Acacia albida* (Fabaceae), *Pyrus syriaca* (Rosaceae), and *Tamarix aphylla* (Tamaricaceae) frequently form extensive stands consisting of a single clone. *Cirsium arvense* and *Centaurea picris* (Asteraceae), *Linaria vulgaris* (Scrophulariaceae), *Convolvulus arvensis* (Convolvulaceae), and *Euphorbia esula* (Euphorbiaceae) produce a primary, positively gravitropic root that forms two different types of lateral roots. Short, thin, ephemeral branch roots function (temporarily) in the absorption of water and minerals. In addition, permanent branch roots are produced that grow horizontally for a considerable distance but take no part in absorption. Eventually, they revert to positive gravitropic movement, formation of branch roots, and normal function in absorption.

The transition from horizontal to vertical growth is associated with developmental activity. Several adventitious buds develop where this developmental change takes place or in the region that just precedes it. In addition, a branch root emerges in the same region. This branch root also grows horizontally and in the same direction as the original horizontal root, acting as its extension and increasing the radial spread of the horizontal root system. This branch root also eventually reverts to negative gravitropic movement, thus repeating the pattern. Adventitious buds formed at or close to the transition from horizontal to vertical movement develop into short, erect rhizomes. When these rhizomes reach the soil surface, they grow into erect aerial shoots, forming branches from the axillary buds of their scale leaves.

In other weed species, adventitious buds are formed at apparently random locations on the horizontal root, not associated with a specific morphogenetic feature. *Apocynum cannabinum* (Apocynaceae) and *Gonolobus laevis* (*Cynanchum laeve*, Asclepiadaceae) produce diagravitropic roots that spread radially and continue moving horizontally indefinitely, producing adventitious buds at random locations along their length.

A strategy that combines two types of diagravitropic roots is exhibited in *Cirsium arvense* (Asteraceae). One type of diagravitropic root develops adventitiously from the subterranean portions of the stem or from the primary root. After moving horizontally a relatively short distance, it reverts to positively gravitropic movement and normal function in absorption. Adventitious buds as well as horizontal branch roots are produced in the immediate proximity of the transition from horizontal to vertical movement. During winter, the autonomous erect plants that develop from these buds eventually become independent, by disintegration of the connecting root.

The other type of horizontal root is a branch of the primary root. It moves forward rapidly, without ever becoming positively gravitropic. It produces numerous adventitious buds at random locations along its entire length, and these eventually develop into erect, autonomous plants with their own adventitious root system. The dual system of horizontal roots in this species thus performs somewhat different roles in occupation and colonization. Whereas the former leads to formation of a colony of new, independent individuals in relatively close proximity to the parent plant, the latter invades and occupies territory by means of an interconnected and thus physiologically integrated clone that is mutually supportive and endowed with a marked capacity for rapid invasive spread.

A strikingly different dual system of diagravitropic roots has been reported in povertyweed (*Iva axillaris*, Asteraceae), a weed native to Canada. A shallow root system is responsible for spread by adventitious bud formation during years with adequate rainfall, while a much deeper lying root system remains virtually inactive. In years with lower rainfall, these deeper lying roots take over the function of spread by growing upward and producing regeneration buds close to the soil surface. They then resume positive plagiotropic growth and once more penetrate to deeper soil layers.

Dispersal

Asexually produced offspring provide the most effective adaptations for dispersal of subterranean propagules of geophytes by migration. Two alternative strategies are employed. One consists of propelling the propagule ahead. A widespread strategy for such migration of the "daughter" bulb, tuber, or corm is by means of stolons. A stolon develops by extensive elongation (intercalary growth) of one or several intermediate internodes (in *Allium* and *Crocus,* respectively, Liliaceae). The stolon acts as a motor organ that propels the young bulb or corm forward through the soil. Movement of the stolon can proceed efficiently only as long as the bulb or corm it subtends are in their early stages of development, sufficiently small to penetrate the soil matrix.

In *Allium ampeloprasum,* the stolon elongates encased in a sheath consisting of tissues growing out of the scale leaf surrounding the plumule. Expansion of the scale leaf to accommodate elongation of its enclosed stolon is probably driven by growth substances produced by activity of the apex of the stolon. The stolon most commonly elongates at right angles to the axis of the parent shoot, probably by a diagravitropic response. This maintains the new plants at the same optimal soil depth as their original parent. In *Poa pratensis* (Poaceae), the stolon apices perform spiral movements (circumnutation) as they grow. These corkscrew movements assist its penetration through the soil matrix.

The other strategy consists of pulling the propagule forward. Most frequently, contractile roots play the major role in such migration (Fig. 18). These roots are initiated at the base of the young daughter bulb or corm at some stage of stolon elongation, moving horizontally, probably by diagravitropic response. The fleshy, proximal part of the contractile root develops acropetally. Its subsequent contraction also proceeds acropetally, and keeps pace with elongation of the stolon. Contraction creates the force pulling the young propagule forward. Its diameter is usually smaller than that of the tunnel vacated by the receding root, facilitating its passage.

The stolon internodes usually start to senesce and disintegrate after their mission has been accomplished and the propagule has reached its destination. Horizontal growth of the stolon maintains the destination of the

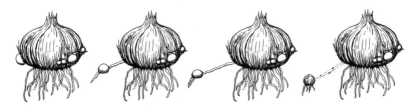

Fig. 18. Some plants use contractile roots to disperse propagules, multiplying and spreading out in the soil.

propagule at the depth of its parent. This general rule usually holds when the depth is optimal. In some instances (such as with *Crocus*), stolons grow at a downward inclination when the parent is displaced too close to the surface, thus relocating the stolons closer to the optimum physiological depth.

Plants such as *Allium neapolitanum* and *Muscari parviflorum* (Liliaceae) have adopted a modified strategy. The bulbils become detached from the parent upon maturation and therefore remain crowded together in the immediate vicinity of the parent until they have sprouted. Once it has sprouted, the bulbil develops a strong, massive contractile root that initially moves perpendicular to the vertical axis of the parent. Subsequently, the contractile root may change its direction toward the physiological depth. In *A. neapolitanum* and *M. parviflorum,* the contractile roots of shallowly planted bulbils grow obliquely downward, and those of deeply planted ones grow obliquely upward. Some of the latter may approach the soil surface then reverse direction until they reach the physiological depth. (Bulbils of *Allium* spp. may migrate distances of 200 mm.)

Seedlings of *Iris sisyrinchium* (Iridaceae) produce a corm at the base of their apical bud. They progressively increase the depth of their corm (up to 200 mm) by means of a strong contractile root that grows vertically down. At the same time, the young corm produces a single axillary regeneration bud. The latter produces a single contractile root that initially moves horizontally, then gradually curves vertically down and may reach a length of 200 mm. As this root contracts, the regeneration bud at its top is relocated laterally as well as vertically down, while developing into a new corm.

An even more aggressive strategy is exhibited by *Oxalis cernua* (and related species of Oxalidaceae). The large parent bulb sprouts to produce an erect shoot that eventually emerges above ground and is equipped with a very thick and long (up to 400–500 mm) contractile root at its base. This root moves horizontally, perpendicular to its shoot, probably by a diagravitropic response. The shoot consists of two developmentally distinct portions. The distal part, which extends out of the bulb, becomes thick and produces numerous adventitious roots. The basal part, which remains within the scales of the bulb, is thin and develops several (usually two to three) regeneration buds in the axils of minute scales. As the horizontal root contracts acropetally (to nearly one-tenth its original length), its contraction is accompanied by a precisely equivalent intercalary elongation of the basal part of the shoot, which thus extends horizontally out of the base of the bulb. As a result, regeneration buds are strung along the horizontal part of the shoot and (more abundantly) along its vertical part in the axils of scale leaves. The latter serve to maintain the new shoots at the physiological depth. The remaining buds are mostly opportunists.

CHAPTER FIVE

Light-Driven Movements

L IGHT IS THE PRIMARY SOURCE of energy for life on earth. Plant leaves are the antennae that collect this energy and convert it to storable and consumable forms of carbohydrate, protein, and fat. Because their life depends on it, plants are adapted to maximizing their photosynthetic activities while also minimizing exposure to hazards of the environments they live in. Thus, they exhibit many movements in response to light, from subcellular movements to behaviors at the organ and whole plant level, including growing toward and away from light, oscillatory leaf movements, and sun tracking.

Certain movements are driven by presence or absence of light, by its irradiance. At the cellular level, the guard cells forming stomata are the most noteworthy example. Plants take up carbon dioxide by diffusion from the atmosphere through their stomata to supply their requirements for photosynthesis. Open stomata allow saturated water vapor to diffuse out of the intercellular spaces within the leaf into the unsaturated atmosphere. Tight management of the exchange of these gases is of vital importance to the plant. This is provided primarily (but not exclusively) by control of the stomatal aperture by light, allowing the stomata (in the majority of plants) to remain open in light and closed in darkness.

Seeds generally germinate underground; in the absence of light, their major source of nutrition is in stored seed reserves. To survive, they have to reach, as rapidly as possible, an environment with photosynthetically active radiation (PAR) that will allow them to start their autotrophic existence before internal supplies run out. In absence of light, their universal strategy starts with the process of etiolation; they accelerate their upward movement out of the soil and into the open air and light by enhancing elongation of their stem, sacrificing radial growth (Fig. 19). The stem of an etiolated seedling is long, thin, and spindly, with little mechanical strength to support its own weight. Upon emergence and exposure to light, the strategy is reversed to de-etiolation: elongation is suppressed, and radial growth is enhanced to produce sturdy stems. Shade-adapted plants (undergrowth) also employ etiolation to accelerate emergence of their shoots from too deep shade. The fluence rate (intensity) of PAR may also be suboptimal for growth, particularly for young seedlings, whose light-harvesting surface is limited. Such plants usually

Fig. 19. Plants etiolate in the dark (*left*) by elongating their stems and not expanding their leaves. When exposed to light, de-etiolation occurs, stem growth is greatly reduced, and leaves are formed and expand.

maximize the interception of PAR by reorienting their apical bud and its developing leaves to face the direction of the predominant flux of light.

In their incessant quest for PAR, most terrestrial plants use the directional signals of sunlight to navigate their light-harvesting organs, reorienting in space or changing conformation for optimizing their utilization of its energy. Some also use these signals to orient their reproductive organs (flowers and inflorescences) and maximize attraction to their biological pollinating vectors.

Many plants exhibit turgor-mediated movements of their mature leaves or leaf parts by means of the pulvinus, the special multicellular motor organ situated in strategic junctions within the leaf, principally at the base of a leaflet or lamina. Pulvinated leaves can perceive the light signal in the pulvinus itself, whereupon the intercepted signal is transduced within the pulvinus to photonastic and/or phototropic responses. Other phototropic leaves perceive the direction of incident light in the lamina. The resulting signal is transmitted to the pulvinus, where it is transduced into responses that move the lamina to face the light.

Photonastic movements are light driven and take place in a predetermined direction, independent of the direction of the light signals, or by transitions between absence and presence of light. Other movements are phototropic, guided by directional light signals such that the direction of movement is tightly coupled to the direction of the light signal. These light-guided movements, exemplified by sun-tracking and other phototropic bendings, are addressed in Chapter 6 (Tracking the Sun).

Signaling with Light

Light, like gravity, provides a basic environmental signal by which plants can control growth, development, and spatial orientation of their shoots, branches, and other appendages. However, light differs from the gravity signal by being extremely complex and variable in its different aspects. Among these are (1) presence or absence of light; (2) transitions between presence and absence; (3) duration (spanning a time-scale over several orders of magnitude); (4) timing (and number of repeats); (5) spectral photon distribution; (6) direction of prevailing light throughout the day or of the incident solar radiation; and (7) exposure of the entire plant, or most of its parts, to diffuse (multilateral), bilateral, or unilateral light (or a specific target may be exposed to an intense microbeam).

Processes driven by light may require light as a signal to get them started. Light can provide the activation energy, acting as a trigger to operate an apparatus that is all primed and ready to continue by means of its own energy. Action of light as a trigger is inductive. Other processes may require continuous input of light as a source of energy; the rate of such light-energized processes frequently corresponds to the irradiance, within limits. Movements of shoot organs as well as movements of specific cells and subcellular organelles are controlled by specific light signals.

Absence or presence of light makes all the differences between an etiolated and a de-etiolated seedling. In dicots, elongation (vertical movement) of the stem (hypocotyl or epicotyl) is suppressed by light; in absence of light, the subapical part of the stem is curved backward in the form of a hook—a result of extreme differential elongation between the concave and convex sectors. In light, this differential is totally reversed, and the base of the hook straightens as it progresses toward maturity. In grass seedlings, elongation of the primary leaf (coleoptile) and of its subtending first internode (mesocotyl) is suppressed entirely by exposure of the tip of the coleoptile to light. At the same time, this induces elongation and expansion of the leaf primordia in the apical bud and their emergence through the coleoptile sheath.

Transitions between absence and presence of light that occur diurnally set the cycle length of the free-running biological clock and change it from its genetically determined circadian length (varying between wide limits, but generally longer than 24 hours) to a diurnal, 24-hour cycle. Return to continuous light or darkness reverts the cycle to its circadian length. The required duration of the light signal may be very short (a few seconds) as in de-etiolation, or considerably longer (hours) as in pulvinar phototropism. The rate of certain developmental processes (including movements) in the following dark period is often influenced by the timing as well as the spectral quality of evening light, the end-of-day (EOD) light signal. Timing, duration, and number of cycles as well as spectral quality of the light are of crucial importance in all processes that are controlled by photoperiodicity. Where photoperiod controls flowering, it initiates movement (elongation) of the flower stalk or reproductive shoot.

Plant Photoreceptors

Three families of photoreceptors control most aspects of light-driven and light-guided movement in plants. Some of them control the rate of movement, and others control the direction of movement; these can operate simultaneously and control both in the same system. New photoreceptors are still being identified. In some instances, more than one photoreceptor has to be activated to facilitate a response, which means that they probably are different downstream components of the transduction chain for that response.

Interactions between Light and Matter

Light is electromagnetic energy. It travels as a wave and acts on a molecule as a discrete unit (quantum) of energy—the photon. The frequency of any specific, monochromatic light is a constant, whereas the velocity of light and thus its wavelength depend on the medium it travels through. The intensity of light is most often expressed with respect to plant biology as the fluence (number of photons) rate (per time) impinging on a given area (per area).

Light is emitted as a monochromatic transverse electromagnetic wave, which means that it is composed of an electric vector (E vector) and a magnetic vector, vibrating as sine waves along a common axis at a specific frequency, in phase and at right angles to each other. The electric component interacts with electrons in a chemical element by causing them to oscillate at its frequency, provided the two are in resonance. In that case, the energy of the photon is absorbed. This means that any given atom or molecule can absorb light photons of one or more specific frequencies (depending on its structure).

Substances that absorb light of specific wavelengths in the visible range (380–700 nm) appear colored and are designated pigments. Their structure consists of separations between electronic energy levels that are sufficiently small to be bridged by photons of light in the visible range. However, the entire biological range is somewhat wider (280–900 nm), including invisible ultraviolet (UV) and infrared (IR). All plant pigments are organic, nonprotein molecules. Photoreceptors for light-driven processes in plants are composed of a close association between a pigment and a protein (apoprotein). The light-absorbing component of the pigment molecule is designated chromophore (carrier of color).

Absorption of light by an atom or a molecule leads to its excitation. Conversely, an excited molecule or atom can emit light of its specific frequency(ies) as the excited electron loses its extra energy (de-excitation). A single molecule can absorb and be excited by a single photon. Therefore, the result of a photochemical reaction is a function of the number of molecules that had been excited as well as of the number of absorbed photons. A photon can excite a molecule only if its energy (as determined by its frequency) is sufficiently high to elevate the electron that it encounters from its low energy ground state to at least the next highest energy level, the excited state (in the higher vacant orbital). Any excess energy is rapidly dissipated, primarily as heat.

The excited state is never permanent, as the high-energy electron has a natural tendency to return to its previous ground state. Loss of energy by de-excitation can follow several alternative pathways, with different rate-constants (velocities). The faster ones obviously take precedence. Emission of the released energy as light of a specific frequency by fluorescence is one pathway.

(continued)

Another pathway is chemical, which transfers the energy to an acceptor molecule. Acquisition of this energy causes the acceptor molecule to undergo a change in its conformation (architecture) that activates it. The chromophore of a photoreceptor is such an acceptor, and the conformational change it undergoes induces a conformational change in its associated protein, which then starts the transduction chain to the final response.

Phytochromes

Phytochromes are red-light photoreceptors; they absorb long wavelengths in the red and far-red regions of the spectrum, with minor responses to shorter wavelength blue light. The photoreceptors comprise a chromophore, which absorbs light, that is bound to an apoprotein, which transduces the light energy to a chemical form.

The phytochrome family contains a number of related members encoded by *PHY* genes. In *Arabidopsis* (Brassicaceae) there are five members of the family PhyA-E, while in other species there are often only two, PhyA and PhyB. PhyA is more sensitive to low fluence rates and is present in high levels in dark-grown and etiolated tissues; PhyB is sensitive to higher fluence rates and is more prevalent than other phytochromes in light-grown tissues. The chromophore of phytochrome is an open-chain tetrapyrrole (phytochromobilin) attached by covalent binding to a protein with enzymatic activity. This molecule exists in two interconvertible forms. One, designated P_r, is generally inactive. This is the form of phytochrome synthesized in the cell; it is stored in seed embryos and is generally the more stable form of phytochrome in the cell. It has peak absorption in the red region of the spectrum (l_{max} 660 nm) and minor peaks in the blue (~380 nm) and in the near-ultraviolet (near-UV ~280 nm) spectral regions. The other molecule, designated P_{fr}, is most often the active form. It has peak absorption in the far-red region (l_{max} 730 nm) and minor peaks in the blue (~400 nm) and in the near-UV (~280 nm) spectral regions. Absorption of red (660 nm) light by P_r converts it into the active form P_{fr}. Absorption by P_{fr} of far-red (730 nm) light converts it into the inactive form P_r. If not converted back to P_r, P_{fr} is unstable and targeted for degradation by the cell.

Conversion between the two forms of phytochrome in either direction involves conformational changes in the chromophore itself and in its apoprotein. There is a certain amount of overlap between the absorption spectra of P_r and P_{fr}, especially in the far-red region. As a result, continuous exposure to far-red light results in cycling between the two forms of phytochrome, a photostationary state and a specific response to light at $l_{max} \sim 710\,nm$. The response is characterized by a requirement for ontinuous irradiation at high levels of irradiance and increases with fluence rate—hence its designation as high-irradiance response.

Blue-Light Photoreceptors

There are two classes of relatively well-studied photoreceptors in plants that mediate plant responses to blue light. These have distinct but sometimes complementary actions, ranging from setting the circadian clock to causing movements both intracellularly and at the organ level.

CRYPTOCHROME. Originally named for its described functions in responses to blue light by the cryptograms (relatively primitive plants such as fungi and slime molds), cryptochrome is a photoreceptor for blue and UV-A light. Two cryptochromes, Cry1 and Cry2, are responsible for the suppression of stem elongation by exposure to nondirectional blue light, but they do not take any part in the phototropic response to unilateral blue light. Cry2 is more sensitive to blue light at low irradiance. Cry1 is a soluble protein that acts as the blue-light photoreceptor for suppressing hypocotyl elongation during de-etiolation (and other aspects of plant growth and development). The protein is encoded by the gene *CRY1* (originally called *HY4*) and is homologous to DNA photolyases, but it has no photolyase activity. Cryptochrome is associated with two chromophores. One is a flavin adenine dinucleotide (FAD), and the other is probably a pterin (methenyltetrahydrofolate).

PHOTOTROPIN. More recently discovered than cryptochrome, phototropin is the prime photoreceptor in growth-mediated phototropic responses to blue light. However, it also controls a variety of nonphototropic movements in response to blue light, such as movement of stomatal guard cells, cytoplasmic streaming (via the cytoskeleton), relocation of chloroplasts, diurnal leaf movements, and rapid suppression of hypocotyl elongation in etiolated seedlings.

Two phototropins, Phot1 and Phot2, have been characterized. Originally called nonphototropic-hypocotyl *(NPH)* genes, *PHOT1* and *PHOT2* encode the apoproteins Phot1 and Phot2 for photoreceptors capable of absorbing UV-A, UV-B, and blue and green light, and regulating all phototropic responses. Phot1 mediates phototropic responses at lower irradiance than Phot2. The absorption spectrum of Phot1 exhibits a major peak at 450 nm, a minor peak at 470 nm, and a shoulder at 425 nm, as well as a broad band around 365 nm (in the UV-A spectrum).

In phototropins, flavin mononucleotide (FMN) is the chromophore. Specific substances that interact with the excited state of flavins inhibit phototropism that depends on blue light. The apoprotein encoded by *PHOT1* is a kinase that phosphorylates itself (autophosphorylation) when the photoreceptor is excited by blue light. The active site of the kinase is located at the C-terminus and contains the amino acids serine and threonine. The N-terminus is distinguished by containing two repeated, closely related sequences (domains) that are typical for a family of similar domains in a large variety of organisms covering the entire biological spectrum; they respond to external signals by light, oxygen, and voltage (the distal sequence, the *LOV1* domain, and the proximal *LOV2* domain, respectively). Both domains bind FMN stoichiometrically. Excitation of the FMN at the *LOV1* or *LOV2* domain by blue light leads to a conformational change in a special helical region of the polypeptide immediately behind them, resulting in activation of the kinase and autophosphorylation.

Collaboration between Photoreceptors

Often, responses of plants to light signals involve collaboration between two or more photoreceptors. Light in the blue spectral region is invariably active in growth-mediated phototropism, but wavelengths longer than 600 nm are also sometimes active. In the fern *Adiantum* (Pteridaceae), coaction is exhibited in phototropism of the protonema as well as in directional movement of chloroplasts in dark-adapted cells of the heart-shaped prothallus. This is also true in suppression of hypocotyl elongation in the de-etiolated castor bean (*Ricinus communis*, Euphorbiaceae). Opening of the apical hook in etiolated seedlings of *Arabidopsis* (Brassicaceae) is stimulated by red as well as by low-irradiance blue light, apparently acting by excitation of phytochrome; it is inductive, reversible by far-red light, and exhibits reciprocity—that is, the response depends on the number of photons perceived (fluence), not on the fluence rate or duration of exposure (Response = Fluence rate × Duration). Hook opening in etiolated dicot seedlings is also enhanced by prolonged exposure to far-red light, apparently acting by means of the high-irradiance response, and by high-fluence blue light, apparently acting by means of a blue-light photoreceptor. High-irradiance response is apparently not involved in the stimulation by red light, presumably because phytochrome B is absent in etiolated tissue.

De-etiolated seedlings of the cucumber (*Cucumis vulgaris*, Cucurbitaceae) exhibit positive phototropic curvature in continuous exposure to unilateral blue light and negative curvature in continuous exposure to unilateral far-red light. De-etiolated seedlings of the *lh* mutant of cucumber, deficient in phytochrome B, exhibit the first positive phototropic response to blue light, but not the negative phototropic response to far-red light. The negative phototropic response to far-red light increases with irradiance. The requirement for continuous exposure to far-red (low-red/ far-red ratio) and the relationship with the level of irradiance suggest that this response may be mediated by the high-irradiance response of phytochrome B.

Vertically growing, young leaves (crozier stage) of the fern *Adiantum cuneatum* (Pteridaceae) exhibit positive phototropic curvature of their midrib in response to unilateral red light as well as in response to blue light. The former is reversible by far-red light, but the latter is not similarly reversible, suggesting joint control by phytochrome and a blue-light receptor. Both red and blue light induce chloroplast movement in gametophytes of *A. capillis-veneris* (Pteridaceae). This dual response is mediated by a photoreceptor gene, *acPHY3*, a member of the phytochrome family. It encodes a unique chromoprotein, phytochrome 3 (Phy3), that exhibits the features of the photoreceptor for phototropism, Phot1. Phy3 is a chimera of phytochrome and phototropin. It exhibits an N-terminal region with sequence homology to the tetrapyrrole-binding region of phytochrome, and a C-terminal region with striking homology to Phot1. This extraordinary chromoprotein may act as a dual sensor that mediates the phototropic responses of *Adiantum* to red as well as to blue light.

Movements Controlling Productivity

Plants depend on the supply of carbon dioxide from the atmosphere for their photosynthetic production of carbohydrates. They facilitate uptake by providing their photosynthetic tissues with an extensive system of intercellular spaces to maximize exposure to the atmosphere. However, exposure of these tissues to the atmosphere results in constant loss of water by evaporation, because their external walls are saturated with water while the atmosphere is permanently unsaturated (with few exceptions). The photosynthetic tissues are well protected against evaporative loss of water by their overlying epidermis, which is coated with water-repellant cuticle and external water-repellant (hydrophobic) wax. This surface is equally impermeable to carbon dioxide, but fortunately plants have evolved an ingenious functional solution to this dilemma.

The epidermis is amply provided with numerous microscopic apertures that are dynamic and capable of regulating the export of water vapor out of the plant as well as the import of carbon dioxide into the plant. These control points are equipped with intricate mechanisms that have the capacity to weigh the requirement for carbon dioxide at any given moment against the hazard of water loss and to respond by changing their dimensions accordingly. These apertures are known as stomata and are universally composed of two symmetrical guard cells around the aperture, surrounded by a cluster of subsidiary cells with an underlying substomatal cavity.

Regulating Import, Export, and Temperature

Changes in stomatal aperture control the flux of gases between the intercellular spaces of the mesophyll and the atmosphere. The requirement for carbon dioxide exists only when light energy is available for its conversion into carbohydrates. A major environmental factor that causes stomatal opening is light. Thus, during daylight the requirement for import of carbon dioxide supersedes the hazard of evaporative water loss, until the latter become critical. When such situation arises, the stomata are capable of sensing hazardously low water vapor pressure in the substomatal cavity and responding to it by reducing their aperture and evaporative water loss. Conversely, when stomata are closed, the availability of carbon dioxide may fall below critical levels. This may seem like a minor shortfall in the supply of carbohydrates; however, the photosynthetic apparatus that is exposed to full light yet unable to convert the intercepted energy into photochemistry is in danger of photodamage, which may become permanent or even terminal (bleaching and necrosis). When such a situation arises, the stomata can sense dangerously low levels of carbon dioxide in their substomatal cavity and respond to it by opening. The guard cells balance the integrated consequences of these three environmental variables and respond optimally to their combination.

Evaporative loss of water through open stomata (transpiration) in daytime may be considered a disadvantage, an unavoidable consequence of the essential requirement for carbon dioxide. However, the water molecules with the highest kinetic energy escape into the atmosphere by evaporation. As a result, the average kinetic energy of the liquid phase and thus its temperature are lowered. Such evaporative cooling is a distinct advantage to the heat balance of the delicate photosynthetic tissues exposed to solar radiation. Furthermore, evaporation of water from leaves produces a tension in the plant's water column traveling through the xylem tubes; this tension provides the major force that moves water and nutrients from the roots up through the plant to the leaves.

There is one important exception to the ubiquitous opening response of stomata to light. An entire group of plants keep their stomata tightly closed throughout the day and open them at night. Most of them form extensive communities in arid habitats and belong to the Cactaceae, Crassulaceae, Agavaceae, Aizoaceae, Euphorbiaceae, Apocynaceae, and Bromeliaceae families, but they are also represented in other families. Fleshy, succulent leaves and/or stems, whose extensive photosynthetic tissue consists of large, thin-walled cells filled with a slimy solution of highly hydrophilic substances, distinguish these plants. They exhibit a special carbon fixation cycle, designated Crassulacean acid metabolism (CAM), because it is characteristic of the entire Crassulaceae family. In these plants, expression of the enzyme phosphoenolpyruvate carboxylase at night causes uptake of carbon dioxide in the dark, resulting in reduced levels of carbon dioxide at the guard cells which in turn stimulate stomatal opening. The carbon dioxide is fixed by carboxylation of phosphoenolpyruvate to oxaloacetate, which is then reduced to malate (or other four-carbon dicarboxylic acids). With the onset of daylight, malate is decarboxylated, releasing carbon dioxide. Elevated carbon dioxide levels at the guard cells cause stomata to close, thus trapping carbon dioxide within the intercellular spaces for fixation in the chloroplasts via the regular, light-driven carbon-assimilation cycle. Clearly, CAM is a highly efficient means for conservation of water under conditions of high evaporative stress during daytime. Water-use efficiency (grams of water lost per grams of carbon dioxide fixed) is five to ten times greater in CAM plants than in other plants. Thus, CAM is prevalent among plant species that inhabit extremely arid environments (such as desert plants and epiphytes); these plants are relatively slow-growing, but evidently survival supersedes rapid growth.

Stomatal guard cells also exhibit circadian cycles of opening and closing, and these are reset to diurnal cycling by light/dark transitions. Stomatal guard cells expand in light and contract in darkness. In this respect, their response to light is opposite to that of the flexor sector of the pulvinar motor cells and corresponds to that of the extensor. Evidently, primary elements of the transduction chains are fundamentally different in the guard cells that are part of the epidermis, and in the motor tissue that is part of the pulvinar cortex.

Stomata open in response to red as well as blue light. However, the response to red light is probably a secondary effect, resulting from lowering the carbon dioxide concentration in the substomatal cavity by photosynthesis. There is little evidence supporting a role for phytochrome in guard cell function. Stomata respond directly to exposure to blue light, even in presence of saturating red light. Protoplasts isolated by removing the cell walls from stomatal guard cells swell in light. The photoreceptors for the stomatal response to blue light are Phot1 and Phot2 and possibly also Cry1. It has been shown that Phot1 when activated by blue light stimulates the proton pump on guard cell membranes, causing them to pump out hydrogen and hyperpolarize the cell, thus drawing in solutes that contribute to water uptake and volume increase.

During periods of drought or related stress, stomatal closure during the day is induced by the hormone abscisic acid, which inhibits proton pumping that depends on blue light. Inhibition of the pump depolarizes the membrane and causes ion and solute leakage, contributing to loss of water and a decrease in cell volume, which closes the stomata. Abscisic acid is synthesized by chloroplasts and accumulates there as long as pH of the stroma (chloroplast interior) is higher than in the surrounding cytosol. This pH differential results from light-driven proton intake into the chloroplast photosynthetic sacs, the grana. Under stress, if photosynthesis is inhibited, protons are released from the grana, and the stroma becomes acidified and releases abscisic acid, which diffuses out of the mesophyll cells and eventually to the guard cells.

Optimizing Harvesting of Light Energy

Plants are autotrophic because they are able to harvest light energy by means of chlorophyll and the photosynthetic apparatus. A highly organized and complex metabolic network makes use of light energy captured by chlorophyll to form an integrated production unit that converts the electromagnetic energy of light into carbohydrates. This activity is housed in the ubiquitous chloroplast of all terrestrial plants, and also present in different forms in algae and certain bacteria. The chlorophyll molecules in the chloroplast are aligned vertically, parallel to each other in tight proximity and attached to the membrane of the extensive, flattened vesicles designated thylakoids. In this configuration, their transition dipole moments maximize the absorption of light normal to the surface of the thylakoid. This configuration exposes all of the chlorophyll molecules to the same flux of incident radiation, thereby maximizing the harvesting of its energy.

Chloroplasts of terrestrial plants are lens shaped, and the thylakoids are aligned parallel to their long axis. Chloroplasts of algae have a variety of shapes, but in all of them the thylakoids lie parallel to their long axis. Thus, the energy harvested by the chloroplast depends on the angle of incidence of light on the flat side of the chloroplast, maximal when perpendicular, minimal when parallel to the incoming radiation. Chlorophyll may absorb radiation in excess of the capacity of the photosynthetic

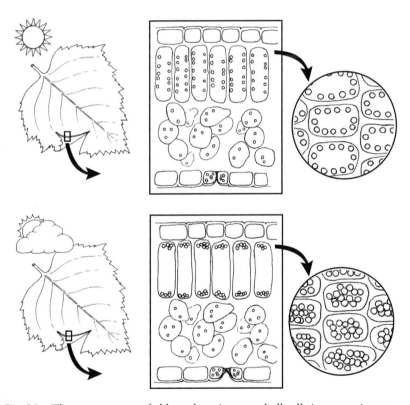

Fig. 20. The arrangement of chloroplasts in mesophyll cells is responsive to light. In bright, sunny conditions (*top*) chloroplasts are spread along the side walls of the palisade cells, as shown in the cross-section (*middle*) and looking down into the cell (*right*) from a surface view. In shade, chloroplasts move to the top and bottom of the palisade cells.

apparatus to utilize it, in which case the excess energy becomes destructive to the light-harvesting machinery. Changing the orientation of the chloroplast with respect to incoming radiation enables control over the flux (Fig. 20).

In terrestrial plants, chloroplasts are located in the photosynthetic tissues of the leaf mesophyll that underlie the epidermis. An extensive system of intercellular spaces provides ample access to carbon dioxide. Chloroplasts move around the periphery of the cell by cytoplasmic streaming, thereby exposing them to light at ever changing angles, but they are capable of aggregating selectively along the anticlinal walls or alternatively along the top periclinal walls of the cell, depending on the level of blue-light radiation they are exposed to. At high levels, the chloroplasts aggregate along anticlinal walls (perpendicular to the plane of the leaf surface), with their thylakoids oriented parallel to incoming radiation to minimize interception (avoidance orientation). This provides protection against the hazards of excessive input of radiant energy. At low levels of incoming radiation, chloroplasts align along the external periclinal walls (parallel with the plane of the leaf surface) with their thylakoids normal to the incoming radiation, thereby maximizing its harvesting (accumulation orientation). In the dark, chloroplasts settle at the bottom periclinal walls. However, the effectiveness of these migrations depends on structural aspects of the mesophyll that are determined by adaptation to the magnitude of prevalent solar radiation.

Leaves of most plants, with the exception of extreme shade-adapted ones, exhibit one or more layers of palisade cells underlying the upper (ventral) epidermis (dorsiventral leaves), and in certain cases a smaller number of layers underlying the lower (dorsal) epidermis (isolateral leaves). Palisade cells are narrow and cylindrical, arranged parallel to each other normal to the leaf surface, along the light beam. In full sunlight, or high PAR, the chloroplasts are aligned along the anticlinal walls of the palisade cells. This avoidance orientation happens also to be the most favorable to light harvesting by the leaf as a whole because the light beam passes along the optically transparent cell axis and on to underlying palisade cells, and thus reaches most of the chloroplasts without loss of energy. Light that reaches the spongy parenchyma below the palisade layers becomes diffuse as a result of internal reflections from the surfaces of the intercellular spaces. Reorientation of the chloroplasts in cells of the spongy parenchyma is therefore absent.

Light for chloroplast movement is perceived at the plasma membrane. This was observed first in the green filamentous alga *Mougeotia* (Zygnemataceae), characterized by a single flat chloroplast along the entire length of the cell. The chloroplast can rotate along its axis, either to face the light or to lie parallel to it. These movements are mediated by phytochrome rather than by phototropin, but they otherwise correspond to the light-harvesting and light-evading movements of chloroplasts in terrestrial plants. Phototransformation of P_r to P_{fr} by red light moves the chloroplast to the light-avoiding orientation within 15 minutes, and this is reversed by re-formation of P_r if followed by exposure to far-red light.

A study of localized irradiation by means of a microbeam eliminated the chloroplast itself as the site of perception and identified the plasma membrane, or its adjacent cytoplasm, as the site of phytochrome for this response. This photoreceptor is arranged anisotropically in a spiral along the cell axis. Use of plane-polarized red and far-red light showed that phototransformation of P_r to P_{fr} by unilateral red light sets up a tetrapolar gradient across the cross-section of the cylindrical cell: P_{fr} at the top and bottom, and P_r at the sectors. Setting up this gradient causes the chloroplast to move its edges from the regions of highest P_{fr} (top and bottom) to those with highest P_r (sectors), thereby rotating the chloroplast to the profile orientation (avoidance). Phototransformation of phytochrome changes the orientation of its transition moment from parallel along the cell surface for P_r to normal to the cell surface for P_{fr}.

In mesophyll cells of terrestrial plants (for instance, *Arabidopsis*) the blue-light photoreceptors Phot1 and Phot2 are responsible for chloroplast movements. The light-harvesting response is activated by either one of the phototropins at low irradiance, while the avoidance response is activated exclusively by Phot2 at high irradiance. This means that Phot2 has a dual role. It can mediate the light-harvesting response at low irradiance, supplemented by Phot1, as well as the avoidance reorientation at high irradiance.

The localization of Phot1 and Phot2 within the cell adjacent to anticlinal or upper periclinal walls is not known except that they are dichroically associated with the plasma membrane. A pulse of blue light may induce the light-harvesting response, expressed by subsequent movement in the dark (within the same cell, but not to its neighbors). Chloroplasts in the light-harvesting orientation move out of a microbeam of high irradiance blue light, but accumulate and even migrate around its periphery. This suggests that the blue light acts by excluding the actin-binding sites for chloroplasts, allowing them to accumulate at the immediate boundary of the illuminated zone.

A key component of the transduction chain for chloroplast movements mediated by blue light appears to be $[Ca^{2+}]_{cyt}$, which also influences the organization of the cytoskeleton on which the chloroplasts may be moored. Phot1 and Phot2 mediate a rapid, albeit transient increase in $[Ca^{2+}]_{cyt}$ from the apoplast in response to exposure to blue light by activation of calcium channels in the plasma membrane. In contrast, Phot2 exclusively mediates import of calcium from intracellular storage.

The helical chloroplasts of the filamentous alga *Spirogyra* (Zygnema-taceae) become tightly coiled where they intercept a microbeam of blue light (l_{max} at 430, 476, and 500 nm, coinciding with those that control migration of chloroplasts in the filamentous alga *Vaucheria*, Vaucheria-ceae; the fern *Selaginella*, Selaginellaceae; the duckweed *Lemna*, Araceae; and the moss *Funaria*, Funariaceae), and their response to centrifugation changes as well. Microbeam exposure of *Vaucheria sessilis* (Vaucheriaceae) to blue light induces the longitudinal cortical fibers of the cytoskeleton to form cross-linkages (localized reticulation), resulting in aggregation of chloroplasts. This response is associated with a light-dependent electric current. Cytoplasmic calcium $[Ca^{2+}]_{cyt}$ is a major factor controlling actin activity, and may therefore also be involved in reorganization of the cytoskeleton. The cytoskeleton may also be involved in control by a blue-light photoreceptor and phytochrome of the movement of chloroplasts in the dark-adapted prothallial cells of the fern *Adiantum* (Pteridaceae) from the anticlinal walls toward a microbeam spot of red or blue light.

Controlling Rate of Movement

Most seeds germinate within the soil, and the seedlings they produce start their life within the soil. Their growth is independent of external resources, except for water, and depends entirely on the food stored within the seed (primarily endosperm) or within their own tissues (primarily cotyledons). Their subsequent survival depends on becoming fully autotrophic before these resources are exhausted.

Accelerating to Autonomy

For a seedling to succeed in establishment, time is of prime importance. The strategy is to move their apical buds as rapidly as possible from darkness to light, to enable their photosynthetic apparatus and organs to become functional. Darkness is sensed as impending starvation, and the response is accelerated elongation of the stem (hypocotyl or epicotyl). To this purpose, the stored resources are diverted into elongation of the stem at the expense of its radial growth. In this state (etiolation), the stem of the etiolated seedling is very spindly, hardly able to support its own weight once it is above ground. However, as the seedling is exposed to light upon emerging into the free atmosphere, etiolation is reversed (de-etiolation): stem elongation is repressed, and radial thickening is promoted. In addition, the cotyledons (in epigeous seedlings) and apical bud, which up to that point had remained small and folded together (as within the seed), unfold, expand, and develop functional green chloroplasts.

The low frequency (long wavelength) part of the biological spectrum light is active in de-etiolation. The photoreceptors for these wavelengths are the phytochromes. Hypocotyl elongation in seedlings of mustard (*Sinapis alba*, Brassicaceae) is strongly inhibited under continuous irradiation with far-red light. This is attributable to the high irradiance response of phytochrome, which establishes the lowest possible photoequilibrium of P_{fr} (<0.03). End-of-day (EOD) irradiation with far-red induces de-etiolation, and this is reversible by subsequent irradiation with red light.

Phytochromes A and B (PhyA and PhyB) are active in etiolation and de-etiolation. Continuous exposure of an etiolated seedling to red light results in conversion of PhyB to P_{fr} and de-etiolation. Continuous exposure to far-red light results in cycling of PhyA to a photostationary state that also leads to de-etiolation, but at the same time it converts PhyB to P_r, which prevents de-etiolation. However, PhyB soon assumes the dominant role in de-etiolation.

Blue light is more active in de-etiolation than red light at equivalent irradiance. Exposure of the hypocotyl of etiolated seedlings to blue light reduces the rate of elongation within minutes or seconds. This fast response is transient (~30 minutes) and is mediated exclusively by Phot1. A gradual, long-lasting decrease in rate of elongation follows, mediated by Cry1. The initial, transient response to blue light, mediated by Phot1, is correlated with a transient increase in $[Ca^{2+}]_{cyt}$. Simultaneous excitation by blue and red light is required for the inhibition of hypocotyl elongation of a de-etiolated castorbean (*Ricinus communis*, Euphorbiaceae), which suggests coaction between phytochrome and blue-light photoreceptors. Apparently, the action of the blue-light photoreceptors depends on the presence of P_{fr}. Hypocotyl elongation in seedlings of the cucumber (*Cucumis sativus*, Cucurbitaceae) is inhibited by blue light, but this inhibition does not take place if the photoequilibrium of phytochrome is minimal (<0.03).

Suppression of growth inhibition by blue light may be associated with reorganization of the cytoskeleton. Curvature of sunflower hypocotyls (and maize coleoptiles) in response to unilateral light (or gravity) is accompanied by reorientation of microtubules at the outer epidermal wall: they move to a transverse orientation along the concave (growth-inhibited) sector, and to a parallel one along the convex (growth-stimulated) sector. Such rearrangement suggests similar arrangement of the cellulose microfibrils in the epidermal cell walls. Growth of individual cells of dark-grown gametophytes of the fern *Ceratopteris richardii* (Pteridaceae) is inhibited when exposed to blue light, and microtubules are reoriented from transverse to longitudinal or oblique with respect to the cell axis. The cytoskeleton, and in particular myosin, may take part in sensory functions and in light signal transduction.

Accelerating to Enhance Light Harvesting

Chloroplasts are highly efficient harvesters of light in the red (600–650 nm) and blue (400–500 nm) spectral regions. As a result, light filtered through a leaf or a continuous canopy of leaves is deficient in the red region and relatively enriched in the far-red region (the red/far-red ratio is ten times lower in such shade). Furthermore, a single chloroplast may absorb as much as 90 percent of the PAR that it intercepts, which means that light transmitted through a second chloroplast may absorb 99 percent. However, in most leaves, a great deal of light is transmitted through the extensive intercellular spaces in the mesophyll, directly, or by internal reflection from the cell walls. In extreme habitats, transmitted PAR may be reduced by two to three orders of magnitude.

Many plants have evolved a strategy that employs the low-red/far-red ratio to evade the consequences of the low PAR. Phytochrome intercepts the far-red-enriched light and starts cycling between P_{fr} and P_r. The photostationary state that is reached accelerates internode elongation at the expense of radial thickening and mimics seedling etiolation. The accelerated growth in height increases the probability of growing beyond the light-limiting conditions under the canopy. Access to sunlight exposes the seedling to blue light, which reverses etiolation. Cell elongation is suppressed with concomitant enhancement of radial thickening.

Deceleration by Blue Light

Blue light suppresses cell elongation and vertical growth of stems in one of two ways. Direct exposure of the growing part of the stem to blue light equally from all sides, or bilaterally, inhibits the rate of cell elongation. Exposure of the growing region of the stem to unilateral blue light results in its positive phototropic curvature by differential elongation, reduced along the exposed sector and enhanced along the opposite, shaded sector. Suppression of cell elongation by blue light is apparently constrained to the cells that actually intercept light; the enhanced growth of the shaded sector is a result of inhibited growth on the other side. Exposure to red light also suppresses elongation of the hypocotyls, but the kinetics are quite different. The response to blue light is much more rapid (seconds) than it is to red light (minutes), as is the time required to recover from the suppression in subsequent darkness.

Several transduction elements have been identified that are associated with the action of blue light in the suppression of elongation by nondirectional light. Rapid inhibition of growth of the cells that intercept blue light is characterized by reduced extensibility of their wall. It is not accompanied by detectable changes in their turgor pressure, and presumably hydraulic conductivity of growing cells is too large to limit their rate of expansion. The initial, rapid inhibition of elongation of cucumber hypocotyls by exposure to blue light is preceded by a large, transient depolarization of the exposed cells caused by blue-light-induced opening of anion channels, allowing these ions to move down their electrochemical gradient. Subsequent inhibition of proton efflux reduces wall loosening, thus inhibiting cell elongation.

Synchronization of Rhythmic Movements

In a constant environment (light and temperature), a variety of shoot organs exhibit autonomous movements on a circadian cycle under control of the biological clock. Under natural conditions, these movements are synchronized by solar timekeeping. Their circadian cycle length is modified to a diurnal one by synchronization with the daily transitions between darkness and light (less frequently by the diurnal temperature fluctuations). Autonomous circadian movements, as well as diurnal ones that are synchronized by light, take place in directions that are predetermined by the bilateral structure and physiology of the organ(s) that execute them, consisting of two opposite sectors of motor tissue along the axis.

The most striking diurnal plant movements are the spectacular ones of flowers and inflorescences. Flowers of water hyacinths (*Eichhornia crassipes*, Pontideriaceae) and species of Onagraceae Cactaceae, Convolvulaceae, and Oxalidaceae exhibit diurnal cycles of closing by inward curvature (hyponasty) of the perianth leaves at their base and opening by reversing the curvature (epinasty). Inflorescences of Asteraceae with ligulate ray florets open and close their floral disc diurnally by similar movements. However, flowers or inflorescences of some plants are open only during certain times during the day. Others are closed in daytime and open at night. These movements are growth-mediated and take place only during development of the flower/inflorescence. Therefore, they may be repeated only a few times or not at all by virtue of the ephemeral active life span of the flowers (Fig. 21).

Diurnal movements of leaves, flowers, and inflorescences are adaptive strategies. Unfolding of leaves in light contributes to light harvesting by their expanded lamina. However, light harvesting is a function of angle of incidence, which undergoes extreme changes in the course of the day, making its interception highly variable. This may have consequences for the degree flexibility required from the photosynthetic apparatus and is ameliorated in those leaves that reorient as the sun moves. Folding leaves at night minimizes their contact with the environment, primarily reducing heat loss to the cold sky by radiant cooling but also reducing evaporative loss. The surface open to attack by pathogens is also minimized. Diurnal unfolding of flowers and inflorescences is of adaptive value to plants pollinated by animals, provided that it coincides with the timing of the diurnal activity of these pollinating vectors. Folding during periods of inactivity has the same protective function as nocturnal folding of leaves.

Circadian movements are synchronized by diurnal light/dark transitions into diurnal movements. It may be assumed that it is the dark/light transition that accomplishes the resetting of the circadian clock. Circadian leaf movements consist of a cycle between a folded and an unfolded con-

Fig. 21. Lotus flowers move their petals to open the flower in the day and close the flower at night.

figuration of the leaf. The unfolded configuration is classified as the photophase, and the folded configuration as the skotophase. Evidently, when synchronization takes place in response to light signals, the response is to nondirectional light signals. The direction of light has no effect on the direction of the movements.

These light signals are perceived in the motor tissue itself, the opposite sectors of the pulvinus. Transitions from light to dark and dark to light result in opposite responses in these two sectors: the extensor expands in light while its opposite flexor contracts during unfolding, and vice versa during folding. This means that when one sector is tightly contracted, its opposite is fully expanded. Perception of the dark-to-light signal leads to contraction of the flexor, and concomitant, passive expansion of the extensor. Therefore, the contracted flexor is less well exposed to the light-to-dark transition, leaving the expanded extensor as the potential site for the photoreceptors for that particular signal.

The location of these sectors is genetically fixed, and may differ in the different leaflets of the same leaf. Movements that take place in autonomous directions, independent of any external signal, are nastic. The fixed position of the two sectors of motor tissue predetermines the direction of movement during folding and unfolding, classifying the movements as photonastic or skotonastic. Furthermore, in some plants the folded configuration is vertically up, in others it is vertically down, and in some it may even be vertically sideways.

The capacity to respond to the dark-to-light signal takes place within a limited timeframe at the start of the photophase (the transition from the circadian folded-to-unfolded configuration) to effectively reset the beginning of the photonastic part of diurnal cycle. In theory, as the leaf is already in the unfolded mode, light is no longer required to maintain unfolding. However, light is still required to reset the remainder of the circadian cycle (nyctinastic) to the diurnal 24-hour cycle. This is accomplished by the light-to-dark transition, which must also take place within a limited time frame of the skotophase. The leaf is maintained in the unfolded configuration by the continued presence of light until the light-to-dark transition takes place. Thus, light resets the circadian clock by acting as an *on* signal and then by maintaining the leaf unfolded until it ends as an *off* signal.

The skotonastic folding starts toward sundown and is therefore strikingly visible to alert observers. The photonastic unfolding, on the other hand, takes place soon after daybreak, when few people are around to observe it and light conditions are not favorable. For these reasons, the popular term for these movements is "sleep movements" or nyctinastic movements. Many pulvinated leguminous leaves exhibit quasi-diaphototropic movements by positive phototropic curvature of their pulvinus in response to its exposure to unilateral light. In this response, any sector of the pulvinus that is exposed to light contracts, exporting water and ions to its opposite sector, which expands as a result. This raises the possibility that the mechanisms of the phototropic and photonastic perception and response are similar, if not identical, but the latter is confined to two predetermined sectors, while the former extends over the entire sheath of motor tissue.

Phytochrome is one photoreceptor responsible for synchronization of the circadian cycle to the diurnal format. The phase of free-running circadian movements in continuous darkness can be shifted by a brief (5-minute) exposure to red light, a response that is fully reversible by an immediate exposure to far-red light. A blue-light photoreceptor is responsible for leaf unfolding at the dark-to-light transition and for maintaining the unfolded configuration. Similar phenomena take place in stomata. The rhythm of stomatal opening in *Phaseolus vulgaris* (Fabaceae) in continuous darkness is phased primarily by the preceding dark-to-light transition, while that of stomatal closure is phased by the light-to-dark transition.

Folding and unfolding of leaves exhibiting turgor-mediated diurnal movements take place when the pulvinus itself is exposed to the corresponding light/dark transition at the appropriate time. In a compound leaf, each leaflet responds and moves independently. Excised pulvini exhibit circadian and diurnal movements in absence of the lamina (but maintaining circadian movements in darkness requires a supply of sugar). Clearly, photoreceptors for the nondirectional light signals are located within the motor cells of the pulvinus. However, opposite volume changes take place in the extensor and flexor in response to the same nondirectional light. Several scenarios have been proposed to account for the opposite responses of the flexor and extensor to the same light signal.

1. On the perception level: Localization of phytochrome and the blue-light photoreceptor in opposite sectors.
2. On the transduction level: Opposite function of ion transporters and channels in opposite sectors.
3. On the rhythmic level: Opposite phasing of the rhythm in the opposite sectors.

However, a much more probable explanation is that these biological motors operate solely by contraction.

Growth-mediated curvature of the motor organ takes place by differential changes in length of opposite sectors, expressed by repression of elongation along one sector and enhancement in its opposite sector. Enhancement is made possible by reallocation of auxin (primarily) and all other resources required for the accelerated growth. Repression of elongation provides the driving force for the response (accelerated elongation) of the opposite sector by passive uptake of resources transported from the repressed sector, but it may also contain elements that are activated by arrival of these resources.

Turgor-mediated curvature of the pulvinus takes place by axial contraction of one sector and concomitant expansion of its opposite sector. Contraction provides the driving force, and the resulting fluxes of solutes and water across the pulvinus drive the expansion by passive uptake but possibly activated by their arrival. This scenario is supported by the fact that in all cases where curvature takes place in response to directional light signals (growth-mediated, pulvinar, and laminar phototropism), the sector that intercepts the light signal is the one that contracts. On this basis, the primary driving force of the pulvinar motor of photonastic/skotonastic leaves is generated in the flexor. The unfolded configuration requires a supply of energy to support it. Folding is a relaxation from the stressed unfolded configuration. The extensor is contracted while the flexor is fully extended to intercept the anticipated light signal.

CHAPTER SIX

Tracking the Sun

THE SPATIAL ORIENTATION of the leaves plays little role in their ability to absorb carbon dioxide for photosynthesis, but it does play a major role in their ability to intercept and absorb photo-synthetically active radiation (PAR). Most leaves are flat and oriented with their adaxial (upper) surface more or less horizontal. In open habitats, the horizontal orientation maximizes the interception of solar radiation throughout the day. However, there are many conditions where the direction of maximal PAR is not vertical, which reduces its interception by leaves displayed horizontally. This could be true for leaves growing next to large opaque objects such as walls, cliffs, or dense vegetation. It is also true for most leaves during the early and later parts of the day, when the sun is rising and setting. This shortcoming has been remedied in many plants by evolving a capacity of reorienting the more active parts of the shoot and/or its leaves to face the direction of maximal incoming PAR.

The direction of movement is tightly coupled to the direction of the light vector, making this a phototropic response. Phototropic reorientation of young, actively growing parts of the shoot, its developing flowers or inflorescences, or its individual expanding leaves is growth mediated and thus is limited by the period of active growth of the tissues. However, where growth is not stationary but advances acropetally, as in the subapical parts of the stem, the phototropic capability moves forward with it. Some plants have extended their capacity for phototropism by evolving a specialized organ for phototropic reorientation of mature leaves. This organ—the pulvinus (see the pulvinus discussion in Chapter 2)—moves uniquely by turgor-mediated phototropism.

Discovery of Phototropism

Phototropism made its scientific debut with a most humble representative of plant parts: the coleoptile of the dark-grown monocotyledonous oat seedling (*Avena sativa* cv. Victory, Poaceae). In the late 1800s, Charles Darwin described the coleoptile curving toward the source of unilateral light. However, it soon turned out that the light signal was perceived in the very tip of the coleoptile, which resulted in curvature by differential growth. This led to the discovery of the first plant-growth regulator, auxin. As a result, the oat coleoptile served for several decades as the model for innumerable studies by plant physiologists working on plant growth substances, with phototropism taking second place. Phototropism became inexorably associated with differential growth, leaving turgor-mediated phototropism far behind, and to a large extent shrouded in mystery.

The fact that seedlings of dicotyledonous plants also exhibited phototropism was noted, but it took a relatively long time to assimilate the difference in their response with that of the oat coleoptile. Whereas in the coleoptile the perception of light is localized at the tip and expresses itself farther down by remote control, the hypocotyl or epicotyl curves only where directly exposed to light. This difference is not fundamental, however. In both cases, exposure to unilateral light affects the activity of auxin transporters, activating preferentially those at the plasma membrane at the side of the cells farthest from the light. The resulting lateral transport of auxin, from the exposed to the shaded side, is ineffective in accelerating growth of cells on the shaded side of the coleoptile tip because the cells at the tip are too immature to respond by elongation. Thus, the auxin is diverted to the lower, actively growing regions of the coleoptile (Fig. 22). In the hypocotyl or epicotyl, the lateral transport of auxin reaches an actively growing tissue and elicits curvature.

It took some more time before it was found that the cotyledons could also act as photoreceptors for differential light exposure and cause a phototropic response in the subtending hypocotyl by remote control. Again, the cotyledon exposed to light transmitted a growth-suppressing signal down the subtending sector of the hypocotyl, presumably by diverting auxin transport to the opposite sector. Eventually, the role of the photoreceptors in the cotyledons was expanded to include the capability of perceiving the direction of oblique light (laminar diaphototropism). A great advance took place when the humble *Arabidopsis* (Brassicaceae) plant made its debut. Its ephemeral life cycle, small size, and, most importantly, small and relatively simple genome soon transformed it into the *Drosophila* of plant physiology, which it has remained to date. Fortuitously, it was highly

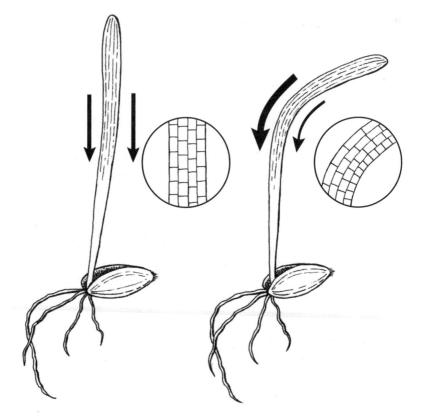

Fig. 22. Phototropism occurs in a monocot seedling. Bending toward the light source is accomplished by more rapid growth, creating larger cells on the side away from the light.

amenable to mutagenesis, and its small size facilitated the mass testing and selection of almost every conceivable mutant, among which those involved in phototropism and in growth were most prominent.

Compromises between Phototropism and Gravitropism

Growth-mediated phototropic curvature in response to direct exposure to unilateral light takes place in young, actively growing parts of the de-etiolated stem. The same or neighboring, parts of the stem also exhibit the universal growth-mediated response to gravity, but the gravitropic response takes place in the basal, most mature part of the curved region. Therefore, phototropic competence may precede gravitropic competence. Separation between the phototropically responsive part of the stem and the gravitropically responsive part that subtends it may be related to the maturation of the starch sheath, the site of graviperception. Negative gravitropic curvature and positive phototropic curvatures of the growing stem take place by differential elongation of the opposite sectors, primarily a result of unilateral growth inhibition in one sector, which may be accompanied by accelerated growth in the opposite sector. The stem integrates the phototropic (and gravitropic) signals acting on it from different directions and responds according to the direction and strength of the resultant signal.

Localization of Perception

Phototropic curvature of the hypocotyl of the sunflower (*Helianthus annuus*, Asteraceae) takes place only when it is directly exposed to unilateral blue light. Shading the apical bud and cotyledons, or removing them, does not affect curvature. The phototropic response of the hypocotyl to exposure of the entire seedling (such as cress, lettuce, mustard, or radish) to unilateral blue light is greater in green than in etiolated seedlings. In etiolated seedlings, the cotyledons and apical bud contribute to the phototropic response of the hypocotyl, as shading them suppresses curvature; but this only delays and reduces curvature in green seedlings. The phototropic response of the hypocotyl to direct exposure to unilateral light depends on its de-etiolation. The requirement for de-etiolation for the phototropic response of the hypocotyl has not been studied.

In de-etiolated (greened) dicotyledonous plants, phototropic control manifests itself under limiting availability of PAR (low irradiance or short duration). Under such conditions, the plant reorients its apical bud and its complement of young leaves to face the predominant direction of light by growth-mediated, positive phototropic curvature of their subtending growing stem. Individual developing leaves exhibit similar responses by curvature of their petiole. Curvature is the result of differential rates of elongation in opposite sectors of the stem, maximal at the irradiated region. Elongation along the exposed sector is suppressed, sometimes totally. Components of the processes of elongation (primarily auxin, water, and osmotically active solutes) are diverted from the suppressed, exposed sector to its opposite shaded sector and contribute to enhancing its elongation.

Moving by Reflection

Plants growing in close proximity to walls, cliffs, and similar vertical, opaque obstacles frequently exhibit curvature of their shoots away from the obstacle. They are responding phototropically to their exposure to differential irradiance: maximal in the direction of prevalent light, minimal in the opposite direction, reflected by the obstacle. In some cases, such phototropic curvature may also take place in diffuse light, in the absence of direct unilateral exposure.

Plants growing in open habitats but close to the edge of other vegetation may also exhibit directional curvature of their shoots away from it, despite being exposed to equal irradiance from all sides (Fig. 23). The

Fig. 23. Plants detect neighboring structures by using the red/far-red receptors, the phytochromes, then move out from under by altering patterns of growth.

intense absorption of red and blue light by chloroplasts of the neighboring vegetation affects not only the spectral quality of the light transmitted through them but also that of the light reflected from them, which is similarly depleted in blue and red, and thus relatively enriched in the far-red region of the spectrum. For this reason, light reflected from dense vegetation is enriched in far-red wavelengths (>700 nm). The shoot intercepts the reflected light unilaterally. Phytochrome in these tissues responds to the low-red/far-red ratio by accelerating elongation on the sector that intercepts such light by the same processes that operate in etiolation. The differential growth-mediated curvature is a phototropic response to differential spectral interception.

Moving Phototropically in Search of Support

Stems of climbing plants (such as *Parthenocissus,* Vitaceae; *Hedera,* Araliaceae; and *Monstera* and *Philodendron* spp., Araceae) exhibit growth-mediated curvature away from the direction of the unilateral light. Such negative phototropism enables them to locate and become appressed to their vertical support (such as walls or trees). Seedlings of the tropical vine *Monstera gigantea* detect and grow in the direction of the trunk of their prospective host tree over a distance exceeding 100 cm. Curvature is in the direction of the darkest sector of the horizon rather than away from the brightest sector (Fig. 24). Furthermore, the magnitude of the response increases with the diameter of the target tree. For these reasons, this response is attributed to skototropism (movement in the direction of darkness), not to negative phototropism. However, when light is too low, the shoot may revert temporarily to positive phototropism in search of light. Some time after the stem starts climbing, it transforms permanently to positive phototropism. This is related to a change in development, as it coincides with a change in the morphology of its new leaves.

In contrast, the prostrate shoot of the fern *Selaginella kraussiana* (Selaginellaceae) grows along the ground by differential phototropism. Its ventral underside exhibits a strong positive phototropic response to direct exposure to white or blue light, while its dorsal upper side exhibits a negative phototropic response that is considerably weaker. These responses are unaffected by the direction of the gravity vector. The direction and magnitude of curvature are similar in fronds that were placed upside down. The negative phototropic response of the dorsal side to overhead light is not unequivocal. It may be accounted for by an indirect positive phototropic response of the lower ventral side to light that filters down to it from above through the frond.

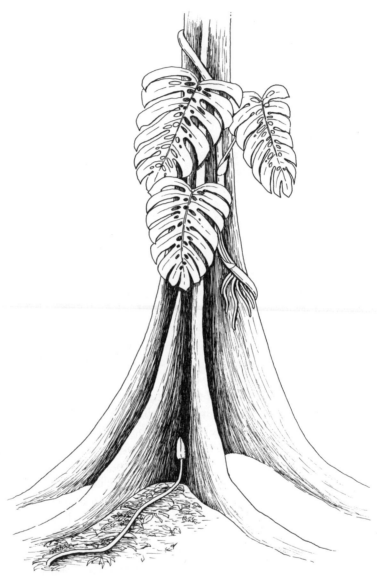

Fig. 24. *Monstera* (Araceae) seedlings detect darkness and grow toward it to find support for growing up as a vine.

Tracking the Sun

A special manifestation of phototropism is sun tracking, or heliotropism. The primary function of the phototropic response is to optimize light harvesting by photosynthetic organs. A secondary function is to optimize the capacity of flowers and inflorescences to attract their pollinating vectors (such as insects and birds). Heliotropism is distinguished from phototropism by a number of fundamental features. Phototropism is a growth-mediated response, generally expressed by curvature of an actively growing tissue, by means of which the plant reorients its apical buds, leaves, flowers, or inflorescences toward unilateral light. Curvature results in a gravitropic response in the same tissues as the phototropic response. The two opposing responses are integrated. Under such conditions, the resultant orientation eventually becomes permanent. In contrast, heliotropism is expressed under full sunlight, and the spatial orientation is continuously controlled by online information from the solar position. The light requirements (irradiance) for the two phenomena differ accordingly. A plant that exhibits a capability for sustained heliotropic movements uses the ever-changing position of the sun to navigate its apical buds, leaves, flowers, or inflorescences according to the solar transit throughout most of the day (Photo 3).

Turgor-Mediated Heliotropism

Leaves of species belonging to diverse taxonomic groups exhibit turgor-mediated diaheliotropism (*dia* = normal to). This response is confined to fully mature leaves. It differs in several aspects from the growth-mediated diaheliotropism of young, actively growing shoot organs previously described. All such leaves are equipped with at least one pulvinus, a specialized structure strategically situated at the juncture between lamina and the petiole, or at the base of the rachis/rachilla in compound leaves. Pulvini may also be situated at the base of the petiole (attachment to the stem). This strategic location enables the pulvinus to move the freely suspended, distal lamina, supported by the stiff tissues at its proximal end.

Two categories of plants exhibiting heliotropic, turgor-mediated leaf movements are recognized. Plants that exhibit laminar phototropism perceive the direction of sunlight in their lamina as vectorial excitation. Their lamina tracks the solar transit throughout the day by positive, remote-controlled phototropic curvature of its subtending pulvinus. The resulting orientation of the lamina is diaheliotropic or plagioheliotropic, depending on whether it remains oriented normal to the sun or is capable of tracking the sun at an oblique angle. Plants that exhibit pulvinar

Photo 3.   Lupine leaves tracking the sun.

phototropism perceive the direction of sunlight in their pulvinus as unilateral excitation and respond by its positive phototropic curvature.
Their laminar orientation with respect to the sun depends entirely on the
positive phototropic response of the pulvinus, changes throughout the
day with progress of the solar transit, and the resulting changes of interception of light by the pulvinus.

LAMINAR PHOTOTROPISM.   Laminar heliotropism is a capability of
fully mature, pulvinated leaves such as those of *Malva* spp. and *Lavatera*
spp. (Malvaceae) and *Tropaeolum* spp. (Tropaeolaceae). These leaves use
their pulvinus as the motor organ to reorient their laminae virtually normal to the position of the sun throughout every clear day with remarkable accuracy and fidelity during most of their active lifetime. The direction of light is not perceived by their pulvinus but rather by their lamina.
The leaf lamina itself must therefore be continuously capable of detecting
the azimuth as well as the elevation angles of the sun. However, its flat
geometry rules out the possibility of an oblique beam intercepted differentially. The normal to the lamina trails the moving oblique beam of
"sunlight" at a small angle, presumably representing the threshold of perception of an oblique beam. (A similar lag characterizes growth-mediated,
diaheliotropic leaf movements in the sunflower.) Directional light is perceived over the entire lamina, not by the pulvinus. Selective shading of
the periphery or center of the lamina (which includes the pulvinus) does
not interfere with the diaheliotropic response.

The lamina reorients by curvature of the subtending pulvinus. Curvature in younger leaves may extend to their still-growing petiole. The resulting curvature is positively phototropic and continuously reorients the
lamina diaheliotropically to face the light throughout most of the day.
The directional signal is transmitted from the lamina in a transduced
form to the pulvinar site of response. Therefore, these leaf movements
are under remote control by vectorial photoexcitation of the lamina, and
they exhibit laminar diaheliotropism.

Perception of directional light. Perception of directional light by the
quasi-flat lamina of diaheliotropic leaves presents a challenge to the clas-
sic concept of phototropism understood as differential interception of
unilateral light. Several studies have addressed possible, alternative
mechanisms by which the leaf lamina may perceive directional light signals without invoking differential interception. Specialized cells in the
upper leaf epidermis of certain leaves act as optical lenses, and it has been
suggested that they contribute to perception of oblique light by focusing
it on specific receptive areas in the underlying cytoplasm.

To test this hypothesis, one half-lamina of *Tropaeolum* (Tropaeolaceae) was covered with water under a thin sheet of mica, assuming that this would eliminate or weaken the epidermal lens effect. The opposite half was left uncovered and dry. When the opposite halves of the lamina were exposed to equivalent but opposite oblique beams, the lamina reoriented toward the oblique beam incident on its dry half, which seemed to support this hypothesis. However, the lamina reoriented toward the light even when the lens effect was similarly eliminated over its entire surface by means of paraffin oil. Moreover, lens-shaped cells are not a common feature in the upper epidermis of diaphototropic leaves.

Another hypothesis invoked local differences in angle of light incidence and the resulting differential pattern of interception of light by the imperfectly flat topography of the laminar surface to account for its capacity to respond to directional light. It was assumed that when the lamina is nonplanar, the differential interception becomes progressively accentuated as the angle of incident light is more oblique and changes with its azimuth angle. On this basis, perception of directional light signals in the lamina of *Lavatera* (Malvaceae) was attributed to increasingly differential interception of PAR by the opposite topography of surfaces on either side of the vein as the azimuth angle of the light beam diverges more from that of the vein. It was assumed that the resulting pattern of assimilate partitioning is therefore differential, and that these differences are somehow transmitted to different sectors of the pulvinar motor tissue, causing the pulvinus to curve. This hypothesis was eventually retracted because the vascular connections did not conform to the prediction.

Vectorial excitation. Directional light, incident on the lamina at an oblique angle, is perceived by the lamina as vectorial excitation. The lamina of *Lavatera cretica* (Malvaceae) forms an incomplete, nearly circular disk with the pulvinus at its center, from which veins diverge palmately in azimuthal directions that differ from each other by ~50°. When one half the lamina is shaded and its opposite half is exposed to light, the lamina reorients when the light is oblique to its surface, not when the light is normal to it; this result is incompatible with the concept of differential interception of light. The rate of response is a function of the vectorial excitation. When vectorial excitation was maintained at a constant level by exposing the lamina continuously to an oblique light beam at a constant angle of incidence moving in an arc along the vertical plane of symmetry of the lamina, as it moved in response to that light, it reoriented at a constant angular velocity exceeding the $15°$ h^{-1} required to track the solar transit.

Sensing solar azimuth and elevation. Perception of vectorial excitation in laminar phototropism is anisotropic. Selective exposure of individual sectors of the lamina resulted in a diaheliotropic response only if the oblique beam was directed along the exposed sector and not at all when it was oriented transversely to it. The only obvious structurally directional tissues in the lamina are those associated with the veins, suggesting that the oblique light is perceived as vectorial excitation in tissues associated with the major veins. Analyses with plane-polarized light suggest that the photoreceptors for vectorial excitation are oriented anisotropically in cell files along the major veins, with their transition moments aligned preferentially parallel to the plane of symmetry of the vein. As a consequence of this anisotropic, orderly orientation of the photoreceptors, vectorial excitation is maximal for the photoreceptors associated with the vein located along the oblique beam. Therefore, these photoreceptors arrays enable the lamina to sense the azimuth of the light vector (solar radiation). The sensory cells have not yet been identified, but they are probably superficial, where scattering of light is minimal.

Vectorial excitation along the plane of symmetry of the midrib at equivalent but opposite angles of incidence results in diametrically opposite responses: increase or decrease in laminar elevation when the beam is directed toward the tip of the vein or its base (tip oriented or base oriented), respectively. When the vectorial excitation is maintained constantly, the opposite responses (laminar reorientation) to either of these excitations take place at constant, albeit different angular velocities. This polarity suggests that the array of photoreceptors along each vein is equally capable of perceiving tip-oriented or base-oriented vectorial excitation.

According to the model that has evolved from these studies, the opposite responses to opposite vectorial excitations could be accounted for by assuming that the photoreceptors that are immobilized at or in the cell membrane, adjacent to transverse walls of each cell along the file, have an opposite orientation of their transition moments at the proximal and distal poles. Directional light signals are therefore perceived as differential excitation of the photoreceptors at opposite poles of each cell along the file. This differential may create a potential gradient between the two poles of each cell as well as across the junction between neighboring cells, which may be expressed as a signal current that is transmitted along the cell file to the pulvinus. The intracellular anisotropic arrangement of the photoreceptors enables the detection of the angle of elevation of the light vector (incident solar radiation) as well as its polarity with respect to the axis of the vein.

The diaheliotropic lamina also senses and tracks solar elevation. The response of the lamina to a sustained vectorial excitation depends on irradiance, as well as on the angle of interception of light by the lamina (maximum between 40° and 50°). This angular dependence rules out the possibility that the photoreceptors are located along and parallel to the longitudinal walls. A more probable location is adjacent to the transverse walls of the cell files along the vein, with their transition moments inclined at some preferential angle to these walls, in addition to being preferentially parallel to the plane of symmetry of the vein.

Vectorial excitation exhibits action dichroism. The integrated net response of leaves of *Lavatera* (Malvaceae) to two simultaneous, diametrically opposite vectorial excitations (transverse to the plane of symmetry of the midvein) with nonpolarized light is proportional to their irradiance ratio. The lamina is capable of discriminating between opposite directional light that differs by as little as 10 percent in irradiance. When the lamina is similarly exposed to two diametrically opposite vectorial excitations by polarized light, with identical irradiance, one with its plane of polarization parallel to the plane of symmetry of the midvein and the other transverse to that plane ($\perp$), it invariably reorients toward the parallel beam. When irradiance of the more effective parallel ($\parallel$) beam is reduced, the rate and direction of laminar reorientation are linearly related to its ratio with respect to the opposite beam ($\perp$). The two beams are balanced at a $\parallel / \perp$ ratio of 0.62. Similar results were obtained with the plagioheliotropic leaf of *Lupinus palaestinus* (Fabaceae). The action dichroism exhibited in the laminar diaheliotropic response supports the concept that vectorial excitation depends on anisotropic orientation of immobilized photoreceptors in the lamina.

The response to vectorial excitation is enhanced by a preceding response to an identical, opposite excitation (by ~40% and ~70% for a base-oriented and tip-oriented excitation, respectively). This suggests a reorganization of the dichroic orientation of the photoreceptors that enhances their perception of the exciting beam of light. Such enhancement would tend to increase the potential gradient along the cell file.

Continuous excitation with blue light is required to drive laminar phototropism as well as to enable pulvinar phototropism, but detailed action spectra are not available. Exposure of the leaf lamina or cotyledon of *Lavatera* to directional red light is equivalent to absence of vectorial excitation (darkness). The lamina reorients normal to its petiole. Leaves of *Malva neglecta* (Malvaceae) respond diaphototropically to blue light, even at relatively low irradiances, but require additional PAR in excess of the light compensation point (the amount of light needed for net carbon dioxide uptake). The rate of response to vectorial excitation increases with the fluence rate of total PAR (400–700 nm), and normal levels of carbon dioxide are required to sustain the response, which suggests a role for photosynthesis-mediated translocation of the directional signal from the leaf to the pulvinus.

From perception to target. The vascular bundles of the major veins coalesce to form the vascular core of the pulvinus, but each maintains its identity and continuity and is associated with a specific sector of the motor tissue. This enables the signal generated by excitation of the photoreceptors along a vein to be transmitted selectively to its subtending (target) sector of motor tissue in the pulvinus, where it is transduced into osmotic activity by controlling transmembrane transport of ions and water. Because the direction of the vectorial light signal (base-oriented or tip-oriented) determines the direction of the (presumptive) potential gradient, it also determines the direction of the pulvinar response. The signal has opposite consequences for its target sector when the excitation is tip-oriented (expansion) or base-oriented (contraction).

In the malvaceous pulvinus, the transduction pathway between vectorial excitation at the laminar site of signal perception and the response at the motor tissue must include exceptional downstream elements: signal transduction in the receptor cells to element(s) that can be transmitted to the site of response along the intervening tissue and probably once more into element(s) that elicit the response in the motor cells (solute transport). It is not known what signal(s) move from lamina to pulvinus in these cases. One possibility is that electric signals move along the vein, differentially expressed by signals from the leaf. Alternatively, laminar signals could alter hydraulic pressure differentially in cells along different sides of the veins, or in some other way could influence transport of a chemical signal to the pulvinus.

PULVINAR PHOTOTROPISM. Field-grown trifoliate legumes exhibit leaf movements that appear to track the sun diaheliotropically in the morning and late afternoon, and paraheliotropically, to varying extents, around midday. It is widely accepted that the primary diaheliotropic movements of such leaves are modified to paraheliotropic when the leaf experiences water stress or stress by supraoptimal PAR and/or temperature around midday. However, with few exceptions, reports of diaheliotropic or paraheliotropic movements have not been supported by time-course studies of the relationship between leaflet orientation and the solar transit throughout the day. The leaf of *Dolichos lablab* (Fabaceae) has been designated paraheliotropic even though its azimuth remains stationary throughout the changes in the solar azimuth. The trifoliate leaf of

Erythrina herbacea (Fabaceae) is down-folded at night. After sunrise, the leaflet laminae incline upward, well beyond the horizontal, and they also rotate axially. This orientation remains virtually unchanged until nightfall, exhibiting no dependence on the changes in the angle of light incidence throughout the day. The leaf exhibits an apparent diaphototropism in the morning and afternoon, and an apparent paraphototropism around midday, despite the fact that its lamina remains stationary.

Pulvinated leaves of many leguminous plants move throughout each day by pulvinar phototropism. Individual leaflets in the trifoliate leaf of *Phaseolus vulgaris* (Fabaceae), when exposed to a transit of a simulated sun in a vertical arc (12 hours at $15°$ h^{-1}) transverse or along its major axis, move independently and exhibit different spatial orientations during different phases of the "solar" transit. Their fidelity of solar tracking throughout the day is considerably less than in the laminar heliotropism of malvaceous leaves under similar conditions. The imperfect fidelity of tracking the sun is reflected in the reduced capacity of the leaf to intercept solar radiation. In laminar phototropism, the directional light is perceived in the lamina as vectorial excitation, and the signal is transmitted to the pulvinar site of response; in leguminous leaves that exhibit pulvinar phototropism, the perception of directional light and the response to it are localized within the pulvinus. This conclusion is based on effects of selective shading of the pulvinus or the lamina from directional light, as well as on the pulvinar responses of debladed leaves to directional light.

Pioneering studies showed that the primary leaf of *Phaseolus multiflorus* and the pinnate leaf of *Robinia pseudoacacia* (Fabaceae) respond to unilateral exposure of the upper (adaxial) or lower (abaxial) surface of the pulvinus to light by increase or decrease of laminar elevation, respectively; lateral exposure results in rotation of the lamina toward the light. In these cases, the pulvinus itself is phototropic. Similar laminar reorientations were observed in the individual leaflets of several trifoliate leguminous species upon unilateral exposure of the entire leaf to adaxial or lateral light. Each of the three leaflets has a different azimuth (by ~90°). Therefore, the resulting movement of the individual leaflets combined different changes in azimuth, elevation, and axial rotation. In the trifoliate leaf of *Phaseolus*, differential exposure of any pulvinus to light results in a positive phototropic curvature. Pulvinar responses become more complex when they combine curvature with torsional rotation, as in lateral exposure, particularly in the two lateral pulvini that are inherently curved in opposite directions. These pulvinar responses to unilateral exposure to light reorient the lamina.

There is thus a clear distinction between perception of light direction in laminar and pulvinar phototropism, and between their heliotropic consequences (as will be described). These aspects are usually overlooked in studies of heliotropic movements of leguminous leaves and do not support the concept of their diaheliotropic or paraheliotropic nature. Leaves exhibiting pulvinar phototropism are heliotropic: their laminar orientation is coupled to the solar transit. However, they do not track the solar transit throughout the day because their laminar orientation depends on perception of direction of light as a unilateral signal, rather than as a vectorial one. They do not maintain a constant orientation with the sun's position, and thus they are neither diaheliotropic nor paraheliotropic.

The pulvinus integrates the phototropic responses of differential light intercepted in its opposite sectors, and its curvature represents their net effect (as described previously in hypocotyls). Selective shading of the adaxial, abaxial, or lateral surfaces of the light-adapted pulvinus (*Glycine max*, Fabaceae) results in laminar reorientation or rotation toward the opposite side. Conversely, phototropic curvature of the pulvinus (*Phaseolus*) by exposure of one of its faces to unilateral light is reversed when its opposite face is simultaneously exposed to similar unilateral light. The magnitude of the phototropic responses may differ in the adaxial and abaxial sectors when exposed to light of the same irradiance, suggesting that the distribution or effectiveness of these photoreceptors, or their capacity for interception of light, may differ in different pulvinar sectors. Most often, each laminar pulvinus in a compound leaf exhibits independent phototropic responses. However, in the trifoliate leaf of *Macroptilium atropurpureum* (Fabaceae), phototropic excitation can be transmitted between primary and secondary pulvini.

Logistics. Information on the mechanism by which continuous unilateral excitation by light causes and maintains the differential activities across the pulvinus is fragmentary. The following sequence of events probably takes place in the course of phototropic curvature of the pulvinus in *Phaseolus*. Continuous exposure of motor cells to blue light of sufficient irradiance excites Phot2, causing efflux of ions accompanied by water from their vacuole into their apoplast volume (cell wall space). Turgor pressure of the illuminated cells decreases, allowing them to contract. A transpulvinar flow of solution changes the ionic environment as well as the water potential of the apoplast in the opposing sector, enhancing the uptake of ions accompanied by water and resulting in expansion.

The hypothesis regarding transport processes into these cells is supported by results showing that protoplasts (isolated from either extensor or flexor) of *Phaseolus* expand in response to increase in external potassium (K^+). In stomatal guard cells, the K^+_{in} channels may function as potassium-sensing valves that open whenever the potassium gradient is inward, allowing potassium uptake only. The activity of potassium channels depends on the voltage of the membrane, which in turn is affected by the activity or inactivity of its proton pump (H^+–ATPase). Phototropic responses of the pulvinus of *Phaseolus* are inhibited by activation of H^+–ATPase as well as by its inactivation, and by blocking potassium channels, which suggest that inactivation of the proton pump in the contracting sector and its continued activity in the opposite, expanding one take part in establishing the turgor differential for the phototropic response of the pulvinus. Continuous adaxial exposure of the primary leaf pulvinus of *Phaseolus* to blue light changes the electric potential difference between the motor tissue in the exposed and shaded sectors from −40 to +20mv, in parallel with the change in laminar elevation. This change in potential difference between the contracting and expanding tissue may be important for the movement of ions between them.

The directional signal for pulvinar phototropism is perceived in the pulvinus itself, probably in the form of a transpulvinar gradient in irradiance. There is no apparent requirement for anisotropic configuration of the photoreceptors. It is therefore likely that they may be located across the motor tissue with random orientation, albeit with increasing concentration toward the periphery. This is not the case with laminar diaphototropism, where the probable location is adjacent to the transverse walls of the cell files along the vein in a highly anisotropic spatial configuration, with their transition moments inclined at some preferential angle to these walls in addition to being preferentially parallel to the plane of symmetry of the vein.

Coexistence of photonastic and phototropic responses. Heliotropic leguminous leaves exhibit both diurnal and circadian movements. Their phototropic and diurnal responses are superimposed. The skotonastic/ photonastic response of the pulvinus and its phototropic response take place by transport of ions, primarily potassium (K^+) and chloride (Cl^-) accompanied by water, out of the contracting sector and into the opposite, expanding one. Similar processes of ion transport are probably involved in both responses. However, different photoreceptor systems control photonastic and phototropic responses of the pulvinus.

1. Opposite volume changes take place in the flexor and extensor in photonastic responses to the *same* light, whereas tropic responses invariably result from contraction of *any* pulvinar sector that is exposed to light and concomitant expansion of its opposite sector. This may be explained by assuming that only the flexor (the contracting sector) is activated by light and its response (contraction and export of solutes and

water across the pulvinus) triggers an indirect opposite response (expansion) in the extensor. However, protoplasts isolated from the extensor of the pulvinus of *Phaseolus* expand in response to increase in irradiance, while those isolated from the flexor contract.

2. The pulvinus responds phototropically to unilateral light from any direction as well as photonastically to a (nondirectional) dark-to-light signal, suggesting that photoreceptors for the phototropic response of the pulvinus are located in every sector. Therefore, the mechanisms of control and response for the phototropic movements exist in every pulvinar sector, and coexist with those of the photonastic movements in the flexor (and/or extensor).

3. The capacity of any sector of the pulvinus to respond phototropically may play a major role in its unfolding at sunrise. In the natural environment, the overhead flux of light predominates, even in absence of direct sunlight, and the lateral fluxes are very similar to each other. Moreover, at sunrise the flexor is expanded and thus intercepts more light than its opposite, contracted extensor. The pulvinus integrates the light signals, unfolding the leaf. The pulvinus starts to respond heliotropically when any of its sectors intercepts light at higher irradiance that disrupts this balance.

Spectral analysis. Pulvinar phototropism is controlled by sustained exposure to relatively high irradiance of blue light, as well as red and far-red light. Pulvinar phototropism in leguminous leaves is driven by continuous unilateral excitation with blue light, with peak activity at 420 and a minor peak at 470–490 nm. The action spectrum for the membrane depolarization induced in pulvinar motor cells of *Phaseolus* by a blue-light pulse exhibits a major peak at 460 nm and minor peaks at 380 and 420 nm. No activity is observed at wavelengths <360 and >520 nm.

The blue and red regions of the spectrum act synergistically in the phototropic response of the pulvinus. Functional differences between red and blue light in the control of pulvinar phototropism become apparent by comparing the pulvinar responses to abaxial or lateral red and blue light, separately and in combination. Exposure of the pulvinus to abaxial, adaxial, or lateral blue light causes positive phototropic responses in each case. Adaxial exposure to blue light increases laminar elevation, and abaxial exposure decreases it, while lateral exposure causes laminar rotation and azimuth to change toward the light, without affecting laminar elevation. In contrast, exposure to adaxial, abaxial, or lateral red light increases laminar elevation. However, laminar rotation, or azimuth are virtually unaffected by lateral red light. The increase in laminar elevation in response to adaxial blue light is enhanced in presence of red light, but the decrease in laminar elevation in response to abaxial blue light is reversed in presence of red light. In contrast, lateral red, which by itself has no effect on laminar azimuth and rotation, enhances the azimuth change but not the rotation caused by lateral blue light.

Lateral blue, which by itself has little or no effect on laminar elevation, enhances the increase in elevation caused by lateral red light. This comparative analysis suggests that red light controls the photonastic unfolding of the pulvinus, whereas blue controls its phototropic responses. These responses coexist in the same tissue, but they are separate and additive. Exposure to red light has the opposite effect on motor cells in the extensor (expansion) and flexor (contraction) sectors of the pulvinus (as expected in photonastic responses of the pulvinus), and exposure to blue light has the same effect (contraction) on either.

The pulvinus of the terminal leaflet lamina of *Phaseolus* responds phototropically to continuous exposure to overhead light by increasing laminar elevation. The kinetics of the response are represented by the initial angular velocity of increase in elevation and by the eventual steady-state elevation. Both parameters increase linearly with irradiance (50–800 μmol m^{-2}s^{-1}). However, the effectiveness of unfiltered white light equals or exceeds the effectiveness of blue light (ll$_{max}$ 450 nm) at equivalent irradiances. The response to blue light (50 μmol m^{-2}s^{-1}) is enhanced progressively by supplementary *red* light (ll$_{max}$ 571 nm) at increasing irradiance (200–800 μmol m^{-2}s^{-1}). Adding red light to blue light is more effective in accelerating the initial angular velocity than adding blue light at equivalent irradiance, whereas adding blue light is more effective in increasing the steady-state laminar elevation. The pulvinus also responds to overhead red (500 μmol m^{-2}s^{-1}) or far-red light (Σ 1,760 μmol m^{-2}s^{-1} >700 nm; 880 μmol m^{-2}s^{-1} between 700 and 800 nm), but at substantially lower rates than in blue light (50 μmol m^{-2}s^{-1}). The kinetics of the response to red light differ qualitatively from those to far-red light. The response to blue is apparently mediated primarily by a blue-light receptor but may involve some absorption by phytochrome, while responses to red or far-red, with and without blue, may be mediated by the high-irradiance response of phytochrome.

Pulvinules of the trifoliate leaf of the kidney bean (*Phaseolus vulgaris*) contain numerous functional chloroplasts that contribute to the pulvinar (photonastic) response to red light and the (phototropic) response to blue light. The pulvinar response to red but not to blue light requires noncyclic electron transport and the resulting generation of adenosine-5'-triphosphate (ATP). Exposure to red light causes membrane hyperpolarization. Depolarization prevents further light-induced changes. The primary effect of blue light is depolarization of the hyperpolarized membrane by inactivating H^+–ATPase, the activity of which depends on a supply of ATP from respiration or photophosphorylation by pulvinar chloroplasts. Unilateral blue light probably induces a decreasing gradient in depolarization of motor cells across the pulvinus, resulting in differential changes in volume.

Blue light activates $[Ca^{2+}]_{cyy}$ and ATP-dependent anion channels in the plasma membrane of mesophyll cells of pea leaves, allowing efflux of chloride. This efflux depolarizes the membrane which in turn activates a proton pump, hydrogen (H^+) efflux, hyperpolarization and uptake of potassium, promoting leaf growth. If the chloride channels also exist in the plasma membrane of pulvinar motor cells, they may be involved in the phototropic pulvinar response (as well as in the photonastic response of its flexor) by enabling the massive efflux of chloride from motor cells upon exposure to blue light.

Motor cells depolarize under conditions of anoxia, and they do not respond to blue light under such conditions. Exposure of the anoxic pulvinus to red light leads to progressive recovery (hyperpolarization) of the membrane potential and of its capacity to depolarize transiently in response to a unilateral blue-light pulse. Under anoxia, unilateral exposure to red light results in negative phototropic curvature, presumably by enhancing hyperpolarization in the exposed side.

Structural considerations. Heliotropic movements in leguminous leaves are functionally distinct from those of malvaceous leaves because they operate by means of pulvinar phototropism. They are most conspicuous in trifoliate species. Leaflets in a trifoliate leaf are oriented at right angles to each other; as a consequence, different sectors of their pulvini are exposed to unilateral light (Photo 4). Therefore, each pulvinus intercepts the same directional light differently and moves independently of the others. Furthermore, the topographical configuration of the pulvinus is different in the terminal leaflet and in each of the two laterals (symmetrical).

Topographical configuration of the pulvinus plays a major role in heliotropism of the leaf. As previously described, the pulvinus of the terminal leaflet is curved upward. In this configuration, its lamina is horizontal when unfolded and obstructs directional light oriented toward its base from reaching the concave (adaxial) face of the pulvinus. The shaded flexor is expanded, leaving the lamina horizontal or down-folded. Phototropic response is suspended until the angle of incidence of the base-oriented light increases sufficiently to be intercepted by the upper surface of the pulvinus. Changes in configuration of the pulvinus in the course of

Photo 4.   Trifoliate bean leaf.

its phototropic response determine the angle and the extent of interception of directional light by the lamina. Such changes may also determine
the penetration of light into the pulvinus. Studies with an optical micro-
probe showed that the configuration of the pulvinus as well as the angle
of light incidence determine the penetration of (blue) light into the adaxial motor tissue in the pulvinus of *Glycine max* (Fabaceae).

Laminar reorientation of trifoliate leaves, as a passive result of pulvinar phototropism, has different consequences for interception of incident
light by the unfolded (light-adapted) horizontal lamina of individual leaflets: the configuration of pulvini of the lateral leaflets is more complex
than that of the terminal leaflet. They are inherently curved upward as
well as rotated to the front (away from the leaf base). This superimposes
torsional rotation over their positive phototropic response to exposure
from any direction. As a result, when the entire trifoliate leaf is exposed

to directional light from the front, rear, or side, the resulting laminar re-orientations reflect these asymmetries: pulvini of the different leaflets exhibit independent, different phototropic responses in accordance with their exposure to such light. Exposure of the terminal pulvinus to vertical light from above or below results in upward or downward curvature, respectively. Corresponding changes in laminar orientation reduce the angle of light incidence (interception decreases). Lateral exposure of the terminal pulvinus to light results in positive curvature of the terminal pulvinus (laminar azimuth changes) combined with torsional axial rotation toward the front. The change in azimuth reorients the lamina less normal to the light (interception decreases), but its concomitant rotation toward the light reorients it more normal to the light (interception increases).

COMBINING LAMINAR AND PULVINAR PHOTOTROPISM. Leaflet movements of *Lupinus* spp. (Fabaceae) are controlled by laminar, as well as by pulvinar phototropism. The compound lamina is palmate (leaflets radiating at different, equally spaced azimuth angles). The entire lamina exhibits diaheliotropic movement (solar tracking) during daytime, as well as nastic down-folding and unfolding.

The heliotropic response depends on perception of vectorial excitation in the basal part of the lamina of each leaflet, not in the pulvinus. At a constant level of vectorial excitation (angle of incidence and level of irradiance maintained constant), the entire compound lamina moves to maintain a normal orientation with the light beam, even when the pulvini of the individual leaflets are selectively shaded. The magnitude of the response diminishes proportionately as the diameter of the shaded area increases to cover a greater segment at the base of the lamina. Shading the pulvinar crown at the center of the lamina does not interfere with the sustained diaheliotropic response, but the leaflets fold down skotonastically.

Conversely, the heliotropic response takes place when vectorial excitation is confined within such a circle but is reduced in proportion to the diminishing diameter of the exposed circle. The heliotropic response also takes place when vectorial excitation is confined to a narrow ring at the base of the leaflets, leaving in shade the pulvinar crown and the rest of the leaflet laminae. Individual leaflets exhibit diaphototropic responses when their pulvinus is unilaterally exposed to light from different directions, after excising their neighbors. A similar role for the laminar base is also suggested by results with selective shading in leaflets of *Phaseolus* and *Macroptilium*.

In the course of solar tracking, the entire compound lamina may exhibit cupping as all leaflets incline upward simultaneously and to the same extent. In the intact leaf of *Lupinus,* the leaflet pulvini are vertical, side by side in a tight crown over the juncture with the petiole, so that their lateral flanks are shaded by their neighbors. As the entire compound lamina moves diaphototropically to remain normal to the sun, only the adaxial surfaces of the pulvini, facing the center of the crown, remain exposed to direct sunlight. When PAR of that light exceeds the photosynthetic capacity of the leaf, all the pulvini exhibit a simultaneous phototropic response and incline upward to the same extent. The entire lamina continues to track the direction of sunlight with its leaflets in the cupped configuration. The combined responses optimize harvesting of PAR by sustained tracking of the sun while continuously adjusting the input of solar energy. The phototropic response of the pulvinus to high PAR modifies the diaheliotropic movements of the individual leaflets to plagioheliotropic.

Growth-Mediated Heliotropism

Apical parts of the shoot may exhibit growth-mediated diaheliotropism. Positive phototropic curvature of coleoptiles of grass seedlings, hypocotyls and epicotyls of dicot seedlings, and actively growing shoots, such as those of the sunflower, is a growth-mediated diaphototropic response of the motor that reorients the apex to face the direction of light. The direction of light may be perceived by photoreceptors associated with the motor itself as differential interception by the exposed and the opposite, shaded sector. Alternatively, leaves may perceive changes in light direction. Cotyledons and leaves clustered around the apical bud differ in their spatial orientation and intercept an oblique light beam at different angles of incidence and different irradiance, resulting in differential growth of the subtending hypocotyls by remote control. However, a diaphototropic response does not necessarily mean a capability to track the solar transit throughout the day, as in solar tracking.

The domestic sunflower (*Helianthus annuus,* Asteraceae) is the most familiar and conspicuous manifestation of such growth-mediated solar tracking (Fig. 25). The apical bud and its cluster of young leaves, and eventually its dish-shaped developing inflorescence, keep moving to remain facing the sun with high fidelity during the course of each day. They do so by growth-mediated positive phototropic curvature of the young, growing part of the subtending stem. Solar tracking is kept up as long as the stem grows, throughout reproductive development, until fruit set. The developing leaves play a role in the diaheliotropic response of the stem, as their excision results in partial loss of the response. However, the site of perception of the solar signal remains unknown.

Fig. 25. Sunflowers (*Helianthus annuus*, Asteraceae) track the sun during the day and reorient toward the east at night.

Similar diaheliotropic responses are exhibited in the numerous inflorescences of the highly branched *Chrozophora tinctoria* (Euphorbiaceae) and *Xanthium strumarium* (Asteraceae) growing in Israel, and wild relatives of *H. annuus* growing in the southwestern United States as well as in flowers of arctic and alpine plants.

Young, developing leaves may exhibit growth-mediated diaheliotropism. Young leaves of the sunflower exhibit diaheliotropic movements even before flower initiation. The laminar orientation of the leaves lags by ~12° behind that of the sun, but maximum "sunrise" and "sunset" orientations of leaves precede sunrise and sunset, respectively, by several minutes. The amplitude of the sunset reorientation of the leaves decreases progressively with flower development. Leaves growing facing sunrise and sunset reorient diaheliotropically by curvature of their petioles and midribs. Those facing north and south do so by axial rotation (torsion). Expanding leaves of *Stachys sylvatica* (Lamiaceae), *Chrozophora tinctoria* (Euphorbiaceae), and *Xanthium strumarium* (Asteraceae) reorient their laminae diaheliotropically by curvature and/or torsion of their petiole and/or of the lamina.

Moving Diaphototropically by Remote Control

De-etiolated dicotyledonous seedlings (cucumber and sunflower) exhibit diaphototropic responses. Direct exposure of their hypocotyl to unilateral blue light results in positive, growth-mediated phototropic curvature that reorients their apical bud and cotyledons to face the direction of the light. However, a shaded hypocotyl also exhibits positive phototropic curvature in response to exposure of the cotyledons and/or young leaves to directional oblique light along their midvein (plane of symmetry) and reorients the apical bud and cotyledons to face the light beam. The motor tissues in the hypocotyl respond to remote control by directional (oblique) light perceived by the cotyledons and apical bud. Perception of the oblique light may be related to the fact that the opposite cotyledons are inclined upward in opposite directions and therefore intercept an oblique beam differentially. The differential growth and positive phototropic curvature may be attributed to the differential supply of growth-regulating substances, resulting from differential interception of light.

This approach is supported by a number of studies, showing that interception of light by one cotyledon inhibits elongation along the subtending flank of its hypocotyl. Seedlings exposed to vertical light with one of the cotyledons shaded curve toward the exposed cotyledon. In absence of one cotyledon, the hypocotyl curves away from the remaining cotyledon, but to a lesser extent in light than in darkness. Diffusates from the hypocotyl on the side with the shaded cotyledon exhibit greater growth-promoting activity.

Young, expanding leaves of *Tropaeolum* spp. (Tropaeolaceae) reorient their lamina normal to an oblique light beam by curvature of their subtending petioles toward the light. Selective exposure of the petiole to unilateral light leads to its positive phototropic curvature, by which its (shaded) lamina reorients to face the light. However, when the lamina is selectively exposed to directional (oblique) light, it reorients to face the light by positive phototropic curvature of its (shaded) petiole. Continuous exposure of the lamina of detached leaves of *Tropaeolum* (Tropaeolaceae) or *Limnanthemum* (Menyanthaceae) to oblique light while it is floating on water results in continuous positive phototropic curvature of its shaded petiole, which forms a coil. Clearly, the positive phototropic curvature of the petiole is remotely controlled by vectorial excitation that is perceived in the lamina. A supply of auxin from the lamina is required for the phototropic curvature of the petiole in response to its direct exposure to unilateral light. Curvature is growth mediated and is greatly reduced in absence of the lamina. This is reversed by replacing the excised lamina with auxin.

Plants maximize light harvesting by forming leaf mosaics, minimizing overlapping by the laminae (Photo 5). Their expanding leaves perceive directional light and exhibit an indirect diaphototropic response by remote control of curvature of their (shaded) petiole. The lamina of *Sparmannia africana* (Tiliaceae) apparently perceives that it is partially shaded and responds by curvature of its petiole, which moves the lamina into full light. The partially shaded lamina apparently produces more auxin in its shaded part and exports this excess to the subtending flank of the petiole to enhance its elongation.

Remote control may account for the diaphototropic response of leaves of shade-tolerant *Hyoscyamus* spp. (Solanaceae) and *Urtica* spp. (Urticaceae) under limiting conditions of PAR, reorienting their lamina to face the direction of prevailing light by curvature and/or torsional rotation of the petiole. Likewise, plants growing near obstacles such as walls or cliffs may exhibit diaphototropism of their developing leaves, and orient them to face the predominant direction of light by remote control. As mentioned early in this chapter, the stem of climbing plants such as ivy (*Hedera*, Araliaceae), Virginia creeper (*Parthenocissus*, Vitaceae), or *Monstera* (Araceae) exhibits negative phototropism by which it grows parallel and in close proximity to opaque vertical supports such as walls and trees. In contrast, the petiole of their expanding leaves reorients their lamina away from the support to face the predominant direction of light, possibly detected by the lamina.

Lactuca serriola and *Silphium* spp. (Asteraceae) are familiarly known as compass plants because the lamina of most of their (cauline) leaves face east–west vertically at maturity (approximately north–south azimuth). In *L. serriola*, the lamina of newly emerged leaves is vertically appressed to the stem, and its azimuth orientation is phyllotactic. As the leaf expands, it reorients its lamina progressively in the course of several consecutive cycles of solar transit across the sky. Expanding leaves with a north–south phyllotactic azimuth rotate their lamina around its midvein, by torsion of the petiole and/or of the lamina itself; those on an east-facing azimuth remain vertical, facing east while those on a west-facing azimuth decline. At maturity, most leaves have their lamina facing either the rising or the setting sun. The nature of the light signals that control this diaphototropic, growth-mediated leaf movement has not been studied.

In mature malvaceous leaves, the vectorial excitation of the lamina leads to positive phototropic curvature of its subtending pulvinus. However, the positive phototropic response can be transmitted beyond the pulvinus. The hypocotyl of seedlings of *Lavatera cretica* (Malvaceae) exhibits positive phototropic curvature in response to its direct exposure to unilateral light, resulting in diaphototropic reorientation of its (shaded) apical bud and cotyledons. The cotyledons are pulvinated and exhibit laminar phototropism. Selective exposure of their lamina to vectorial excitation reorients it normal to the light by pulvinar curvature, while

Photo 5. Leaf canopy mosaic.

the (shaded) hypocotyl remains upright. When the entire seedling is exposed to directional light, positive phototropic curvature of the hypocotyl satisfies the diaphototropic requirement of the cotyledons without pulvinar intervention. In explants of seedlings (hypocotyl and a single cotyledon), the (shaded) hypocotyl curves away from the remaining cotyledon when its lamina is in darkness or exposed to vertical light. Exposure of the lamina alone to continuous vectorial excitation with blue, but not with red, light enhances or reverses the curvature of the (shaded) hypocotyl (as well as of the shaded petiole), depending on the direction of vectorial excitation: enhanced curvature away from the cotyledon when the excitation is tip-oriented, and toward the cotyledon when it is base-oriented. The signal generated by the vectorial excitation of the lamina is transmitted to the hypocotyl, causing it to curve by differential growth, possibly by driving fluxes of solutes and water from its concave to its convex flank, as in turgor-mediated pulvinar curvature.

Modification by Stress

Pulvinar phototropism of leguminous leaves may be strongly modified when the leaf experiences a variety of stresses. Suboptimal leaf water potential and supraoptimal leaf temperature are the major stresses that modify heliotropic movements of pulvinated leaves. Steady-state laminar elevation of light-adapted primary leaves of *Phaseolus* is greater at higher irradiance and increases progressively with leaf water stress, expressed by increasingly negative leaf water potential. Heliotropic leaf movements of light-adapted leaves of *Phaseolus* are progressively modified by increase in ambient air temperature, under constant atmospheric humidity and carbon dioxide, or mesophyll carbon dioxide. Selective exposure to radiant heat showed that heliotropic leaf movements are modified by supraoptimal temperature of pulvinar tissue, and they are not mediated by effects of temperature on leaf water potential, transpiration, or stomatal conductance. Furthermore, phototropic curvature of excised pulvini of the bean increases progressively with temperature. Modification of such leaf movements by temperature and water stresses takes place only in light. It may therefore be assumed that these stresses enhance the transport of ions and water from the exposed to the shaded sector of the pulvinus in the course of its phototropic response. This enhancement may take place by means of increases in levels of abscisic acid (ABA) in the exposed sector, mediated by stress.

The stress hormone ABA inhibits circadian leaf movements of *Oxalis regnellii* (Oxalidaceae), possibly by reducing membrane permeability, and it may play a similar role in the exposed sector of the phototropic pulvinus and in stomatal guard cells. ABA initiates the transduction chain leading to contraction of the guard cells by increasing the level of free calcium in the cytosol. High irradiance intercepted by the pulvinus transforms the (apparent) diaheliotropic orientation of the lamina into an (apparent) paraheliotropic one. However, the steady-state laminar elevation is a function of the (adaxial) irradiance intercepted by its pulvinus. It is therefore more likely that the observed change from diaheliotropic to plagioheliotropic orientation upon exposure to high irradiance is only a manifestation of the phototropic response of the pulvinus, and that the lamina is normal to the light only within some critical, intermediate range of irradiance then diverges progressively as irradiance increases or decreases beyond this range.

The leaflet laminae of the palmately compound leaf of *Lupinus* spp. (Fabaceae) respond to a variety of stresses simultaneously. In the absence of stress, individual leaflets of *L. arizonicus* move diaheliotropically throughout the day. As water stress develops (expressed as reduced leaf water potential) around midday, movement of the leaflets becomes plagioheliotropic (cupping) and tracks the sun at an acute angle. A similar modification takes place in response supraoptimal irradiance, which modifies its diaphototropic response to a plagioheliotropic one.

The flux of solar radiation intercepted by most terrestrial plants may exceed the capacity of their leaves to lose thermal energy through reradiation, convection, and evaporative cooling. Furthermore, the flux of PAR that is intercepted by the chloroplasts may exceed their capacity to use it for photosynthesis. This may lead to photoinhibition as well as to eventual destruction of the photosynthetic apparatus. The hazards of excess solar radiation are reduced primarily, but not entirely, by mechanisms within the chloroplast for harmless dissipation of the excess light energy it absorbs. These mechanisms are apparently sufficient to protect certain plants that reorient the laminae of their mature leaves to maximize their interception of direct solar radiation throughout each day. Other plants minimize such hazards by evasion, moving their chloroplasts and/or leaves to orientations that minimize the interception of solar radiation. Some achieve evasion by changing their conformation.

Diurnal Heliotropism

Plants organs exhibiting diaheliotropism end each day facing the direction of sunset. Were they to remain in this orientation, they would be obliged to waste precious time at the following sunrise to reorient toward the rising sun. These plants avoid this difficulty by a strategy of diurnal heliotropism. Although the mechanisms and motors are quite different in growth-mediated and turgor-mediated diaphototropic responses, both exhibit a strikingly similar nocturnal reorientation.

A very familiar and hard to miss example is provided by the diurnal movements of the apical bud of the sunflower (*Helianthus annuus*, Asteraceae) and its developing inflorescence that are repeated through successive 24-hour cycles until fruit set. Throughout the day, they track the sun with high fidelity. At the end of the day, they are oriented toward the direction of sunset. Sometime after sunset, they start reorienting in the opposite direction and end (well before daybreak) by facing in the anticipated direction of sunrise. However, the predaybreak orientation toward the rising sun is only approximate and is readjusted by tracking the sun after sunrise. Nocturnal reorientation is more rapid ($\sim26°$ h^{-1}) than tracking the sun (at $\sim15°$ h^{-1}) during the day. The direction of nocturnal reorientation remains unchanged, even when the preceding day was overcast, and even three to four days after the (potted) plant was rotated $180°$ around its axis. Therefore, it appears to be dictated by the preceding solar tracking.

Malvaceous leaves exhibit a similar nocturnal reorientation. After sunset, mature leaves of *Malva neglecta* and *Lavatera cretica* start to reorient their sunset-facing lamina to end facing the direction of the anticipated sunrise, several hours before it occurs. Plants that are rotated by $180°$ at sunset persist in their original direction of nocturnal reorientation to face the preceding sunrise. Facing directly opposite to the new sunrise, their adaxial (upper) face is not exposed to the sun for several hours, during which they are unable to start their daytime heliotropic movement in the absence of directional information. As a result, the leaves resume normal nocturnal reorientation to face the "new" sunrise only after the plants adapted to their new position during several cycles. The direction of nocturnal reorientation appears to be predetermined by that of the preceding sunrise and may therefore be considered as the nocturnal phase of diaheliotropism. The related *Lavatera cretica* exhibited similar diurnal movements under field conditions as well as during several consecutive cycles under simulated conditions of diurnal diaheliotropism by means of a solar simulator.

Three phases may be distinguished in nocturnal reorientation of the lamina: (1) pulvinar relaxation from the strained, sunset-facing configuration (duration of this phase depending on the extent of laminar displacement required); (2) pulvinar equilibrium (time measuring); and (3) reorientation to face sunrise. Cotyledons also exhibit diurnal heliotropism under simulated conditions, but acquire the capacity for nocturnal reorientation after experiencing three to four cycles of diaheliotropic movements.

These plants exhibit diurnal heliotropism by complementing their day-time solar tracking with remarkably precise nocturnal reorientation. In navigational terms, malvaceous species guide their leaf laminae, and sunflower species guide the orientation of the apical bud complex (and subsequently the inflorescence) by the solar transit during the day and by an automatic pilot during the night.

CHAPTER SEVEN

Prey-Driven Movements

ERTAIN PLANTS are incapable of autotrophic life because they lack necessary equipment such as chloroplasts. These plants, classified as holoparasites, obtain their entire organic nutrition by invading autotrophic plants and tapping into the vascular system that transports assimilates from their leaves. Other plants possess the necessary equipment for producing their own organic assimilates but lack the equipment for absorbing water and solutes from the soil. These plants, classified as hemiparasites, obtain their water and essential minerals by invading other autotrophic plants and tapping into their water conducting vascular system. Another group of plants have taken refuge in nitrogen-deficient habitats and have developed an appetite for nitrogen-rich food in the form of ready-made amino acids to supplement their uninteresting diet of homemade assimilates. Using ready-made amino acids is a great savings in metabolic energy, and these plants obtain their gourmet food from animal sources—primarily insects that they capture by a variety of means then kill and digest, absorbing the digested protein content of their corpses. These plants are classified as carnivorous (insectivorous).

Parasitic Plants

Parasitic plants must first locate their favorite prey. Different strategies are employed by different species. The mistletoes *Viscum album* and *V. cruciatum* (Santalaceae) and *Loranthus acaciae* and *L. europaeus* (Loranthaceae) are hemiparasites that find their target victim by producing attractive, tasty fruits with bright colors. The succulent fruits are favorites with birds, who eat the fleshy coat but find it quite difficult to get rid of the indigestible, hard seed that remains stuck to their beaks. When they wipe their beaks against a branch, they leave the seed deposited there, and the seed may germinate. The tissue forming a root-like structure at the base of the seedling shoot rapidly invades the host tissue and grows into a massive haustorium. The latter moves by growing toward the water-conducting tissue of the host, to which it bonds itself as tightly as a graft. Presumably, the haustorium is guided toward its target tissue by a chemotropic response to some (hitherto unidentified) substance emanating from the target. The (green) seedling shoot starts developing only after the grafting is accomplished.

A different strategy is employed by dodder (*Cuscuta* spp., Convolvulaceae) (Fig. 26). This holoparasite produces innumerable seeds that

Fig. 26. The parasitic dodder seedling (*Cuscuta*, Convolvulaceae) "sniffs" its host, grows toward it, then coils around it for support before merging tissues and withdrawing nutrients.

eventually germinate indiscriminately. The seed carries a sufficient supply of nutrients to sustain early growth of the seedling for a considerable time. The seedling uses this delay period to elongate rapidly and perform extensive circumnutation movements. Recent evidence shows that dodder seedlings detect (smell) volatile compounds produced by neighboring plants and tend to grow preferentially toward more suitable hosts; thus, dodder readily invades tomato plants but is not as attracted to wheat, which it is much less able to penetrate. Contact transforms the random nutation movements of the shoot into a thigmotropic response that establishes permanent differential elongation, inhibited along the flank in contact with the host shoot and enhanced along the opposite flank, forming tight coils around the host. Wherever tight contact is established, the parasite invades the host, producing a haustorium that finds its way to the conducting tissues for assimilates as well as water. In absence of a host, the seedling is capable for a time of moving along the soil surface, feeding its apical elongation with nutrients salvaged from the death of its older, distal part.

Unlike the strategies employed by parasitic plants that invade their victims through their shoots, holoparasitic plants such as broomrape (*Orobanche* spp., Orobanchaceae) also produce innumerable seeds that are widely distributed, but their seeds are stimulated to germinate only by special substances that the roots of their victims exude into their soil environment. Roots of the resulting seedlings randomly locate and invade a host root. Invasion is followed by growth and development of the parasitic shoot, which forms a subterranean bulbil and a spectacular erect flowering stalk above ground.

Carnivorous Plants

Carnivorous plants are at a disadvantage because they are rooted in one place but their prey are fully mobile, at liberty to move rapidly in any direction. These plants have developed strategies for trapping their victims, often luring them into the traps by a variety of means—most commonly by sight and/or smell. The traps are usually highly elaborate modifications of specialized leaves.

Passive Traps

In certain carnivorous plants, the traps are totally passive. The Sarraceniaceae (pitcherplant) family uses a highly attractive, deep, pitcher-like container formed by the lamina and filled with digestive enzymes of proteins that are secreted by specialized glands. The prey is lured to enter the particularly attractive neck of the pitcher, which is equipped inside with a highly slippery coat. The victim then either slips down to drown in the enzyme soup and be digested by it, or it finds its exit blocked by long, stiff, slippery hairs pointing one way—down into the enzyme soup. Such traps are stationary and do not involve movements.

Traps evolved by other carnivorous plants, such as *Pinguicula grandiflora* (Lentibulariaceae), are considerably more elaborate, involving modification of the surface of the lamina of the otherwise green, photosynthetic leaves produced as rosettes during summer only. The margins of the elliptical leaves are turned upward, forming a dish. The traps consist of a more or less uniform field of erect, stalked, glandular hairs that sprout from the surface of the lamina and exude a droplet of very sticky mucilage at their tip. The glistening field of droplets, which are frequently pigmented, attracts insects to land on the glandular hairs and become irrevocably enmeshed in their sticky mucilage. The trapped insect's movements to free itself excite additional glands and stimulate the exudation of more mucilage, which eventually engulfs and smothers the victim. Other glands are sessile on the laminar surface and produce hydrolytic enzymes (ribonuclease, esterase, acid phosphatase, amylase, and protease) that digest the entire content of the victim, leaving an empty chitin shell.

Struggles by larger trapped victims initiate movements of the leaf, increasing the efficiency of the trapping mechanism. The stalked glands that are involved in trapping rapidly lose their turgor and collapse onto the laminar surface. At the same time, the epidermal cells surrounding the collapsed gland also lose turgor, causing the lamina edges to curve upward in the form of a bowl. The slow movement of the upturned margins of the lamina (lasting about two days) deepens the bowl. The trapped insect comes in contact with more mucilage from stalked glands that bend inward and with the hydrolytic enzymes secreted by sessile glands on the laminar surface, which accumulate as a pool forms in the bowl.

Like *Pinguicula*, the sundew (*Drosera* spp., Droseraceae) traps its prey by means of sticky mucilage, exuded by glands raised above the laminar surface on top of long stalks. However, the trap is considerably more elaborate and exhibits larger scale movements. Leaves of different species exhibit a great diversity in size (1.5 mm to 60 cm) and shape (simple or divided), organized either as rosettes or along the erect stem. In *D. rotundifolia*, the upper surface of the leaf lamina carries numerous erect, flexible tentacles, whose flattened top is crowned with a gland carrying a droplet of clear, sticky mucilage that becomes reddish and glistens in sunlight. The amount of mucilage is precisely measured, just sufficient to form a droplet on the top. The field of glistening, colored droplets creates a spectacular vision that attracts prey to land on the tentacles where they are trapped by the mucilage.

In addition to the mucilage, the exudate contains digestive enzymes (peroxidase, esterase, proteases, and acid phosphatase) that break down and digest the animal. The gland eventually absorbs the products of digestion. The tentacles are arranged in concentric circles around the center of the lamina. Their morphology and function change progressively from the center of the lamina to its outer (marginal) circle. The length of the tentacles increases progressively from the center to the margins. They are inclined outward at angles that become progressively more acute toward the periphery. The central ones are virtually erect, while the marginal ones extend out in the plane of the lamina. In addition to their role in producing mucilage as the trapping agent and their sensory capabilities, the function of the tentacles changes progressively from the center to the periphery of the lamina. The former act primarily in a digestive capacity and are virtually immobile. In contrast, the peripheral tentacles respond to movements of prey that they themselves trap, as well as to signals emanating from other tentacles that have struggling prey trapped in their mucilage. All the tentacles are capable of sensing the presence of a trapped insect and transmitting signals to that effect to their neighbors. Generating a signal does not require struggles of live prey and can be simulated by mechanical perturbation using an inanimate object. Mechanical excitation, by living trapped prey or by simulation, is perceived just below the flattened glandular head, where the diameter of the tentacle is smallest, facilitating its deformation.

The distance traveled by the signal depends on the size of the prey and the magnitude of its struggles to free itself from the mucilage. The primary response is seismonastic: rapid movement by curvature of the perturbed tentacle along its plane of symmetry, toward the center of the lamina. Curvature takes place by progressive differential growth, greater along the external (dorsal) flank than on opposite (ventral) flanks of the tentacle, from the base toward the tip. It may start within several seconds or minutes of stimulation by prey (or mechanical means), enabling rapid immobilization of larger insects. Tentacles of *D. drummondii* curve ~180° in less than 1 minute. The primary seismonastic response is fully reversed by compensatory growth after digestion of the prey had been completed. (Mechanical stimulation in absence of a trapped insect allows the curvature to be fully reversed within 24 hours.) Movement is growth mediated, and any single tentacle is limited to a few performances. This limitation does not affect the responses of other tentacles.

As a result of these nastic movements, the tentacles converge toward the center of the lamina, the primary site of the digestive tentacles, causing the prey to become submerged and smothered in a pool of digestive mucilage. Intensive mechanical stimulation produced by struggles of large trapped prey produces a different response. Neighboring, free tentacles now exhibit curvature toward the source of the signal, which is the focus of the intense perturbation by the prey. They now add their mucilage to completely smother it. Further movement of the adjacent tentacles transports the victim toward the center of the lamina, where digestive activity is maximal.

Concurrently, the entire lamina curves upward, forming a mucilage-filled bowl that is soon to become the focus of digestion. These movements take place in the direction of the signal and are clearly seismotropic. The seismonastic response results from a signal traveling down the tentacle from the sensory apparatus at its top, activating an inherent differential growth on opposite sides of the plane of symmetry of the tentacle. In contrast, the seismotropic response results from a signal generated by neighboring excited tentacles reaching the base of the target tentacle, then traveling up along it. The signal forms a gradient along the direction from which it arrives, thus determining the direction of the plane of the differential elongation and of the resulting tropic curvature of the tentacle.

Primed Traps with Mechanical Triggers

The movements generated in the traps of *Pinguicula* and *Drosera* take place only after the prey had been trapped by the mucilage and are a direct result of stimulation by contact with the trapped prey and by its struggles to free itself. Traps of other carnivorous plants are primed for an immediate, rapid response and are triggered by mechanical perturbation caused by contact with the prey. Perception of such perturbation by the trap's sensory mechanism operates a trigger that causes an immediate release of preexisting tension in the trap, causing it to close around the prey. Digestive processes follow, and when these are completed by absorption, the tension is reinstated, resetting the trap to its primed condition. To be effective in trapping an actively mobile prey, a triggered response must be quite rapid. This is only possible in living tissue of a plant if the trigger sets off a turgor-mediated response.

The trap of the Venus' flytrap (*Dionaea muscipula, Aldrovanda,* and related species of the Droseraceae) is formed by metamorphosis of the lamina of its specialized leaves (Photo 6). It consists of two almost symmetrical lobes on opposite sides of the midrib. The edge of each lobe is equipped with a single row of equally spaced teeth (14 to 21)—long, stiff bristles that face upward nearly perpendicular to the laminar surface in the form of a rake. The immediate response of the trap to the appropriate excitation is a very rapid movement of the two lobes toward each other, with the midrib as the operative hinge. The teeth on the opposite margins are spaced precisely to interlock as the opposite lobes move toward each other, forming a cage and thus trapping the prey.

In each lobe, a thinner peripheral band surrounds a thicker central region. The upper (adaxial) surface of the peripheral band carries numerous sessile glands (flush with the surface of the epidermis) that secrete a solution of carbohydrates, presumably nectar, that apparently attract the prey. In the related species *Aldrovanda*, the glands of the peripheral band are quadrifid (with four cells radiating outward in the form of a narrow X). This species is aquatic, and similar quadrifid glands are characteristic of the aquatic but unrelated genus *Utricularia* (Lentibulariaceae). Presumably they function in pumping water out of the trap.

The central region of each lobe carries three widely spaced, long, thin, conical bristles (numbers may differ in other species) that end in a

Photo 6.   Venus' flytrap.

swollen joint above a distinct constricted segment. The bristles are stiff along their entire length except for the constricted region at their base that remains flexible. Mechanical perturbation of the bristle that is sufficient to cause the flexible joint to bend and distort acts as a trigger for the response. Cells within the flexible joint exhibit an accumulation of tightly packed, concentric endoplasmic reticulum at their upper and lower poles. These and additional anatomical features are probably components of the sensory mechanism that is activated when the cells are distorted by flexing the bristles, and start the transduction chain to the rapid movement of the trap. These bristles are strategically situated to detect presence of prey. However, (in *Dionaea*, but not in *Aldrovanda*) the presence of prey must be verified and confirmed either by repeated stimulations of one such sensitive bristle or by stimulations of more than one bristle within about 20 to 40 seconds of each other. Only then is the trigger activated, resulting in the rapid movement of the opposite lobes toward each other. In addition to the touch-sensitive bristles, the upper (adaxial) surface of the central zone caries numerous glands that produce the mixture of digestive enzymes when stimulated by chemical secretions of trapped and crushed prey.

Rapid movement (closure) of the trap is turgor mediated: cells of the upper epidermis along the midrib rapidly lose turgor by a process probably involving bulk transport of water and ions—primarily potassium and chlorine—from these cells to the lumen of the trap. Excess turgor in the cells of the lower epidermis underlying the midrib and of their internal neighbors forces the lobes to move upward toward each other, with the midrib as a hinge. The rapid phase lasts ~1 second. It does not require excitation by movements of living prey and may take place in response to successive mechanical perturbations of the sensitive bristles. In absence of prey, the rapid movement elicited by mechanical perturbation is subsequently reversed; the cells along the upper epidermis regain turgor, moving the lobes apart to their original orientation. Being turgor mediated, the rapid movement can be repeated many times with repeated mechanical excitation, although opening requires energy and may be slow on cool or shady days. The effectiveness of the repeated excitations in triggering closure diminishes as the interval between them increases. Mechanical excitations repeated at longer intervals (1 minute) result in progressive closure of the lobes, the magnitude of each stage exceeding that of its predecessor.

The rapid phase ends with the marginal teeth of the opposite lobes interlocking to form a cage (trap closure); the lobes themselves are concave, forming a space enclosing the prey. If the prey is live and moving, the capture movements of the leaves are followed by slower movements (narrowing phase) of the lobes toward each other. The thin peripheral band of each lobe moves to bend outward along its border with the thick central region, eventually lying flush against each other to form a tight seal around the central part. This forms the digestive cavity in which the prey is imprisoned and eventually digested. The outward movement of the peripheral region during the narrowing phase causes movement of the marginal teeth to straighten, disengage, and lie parallel to their opposites in the plane of the lobe.

Movements during the narrowing phase are growth mediated, as is the recovery during reopening from this phase. This places a limit on the number of times this narrowing activity can be repeated. Presence and activities of the prey induce secretion of digestive enzymes into the tightly sealed digestive cavity. The trap remains in this tightly sealed condition until digestion is completed, which may last a week or longer. Reopening the trap follows the end of digestion and is heralded by movement of the marginal teeth to their original orientation, perpendicular to the lobe.

There is general agreement that the aquatic bladderwort (*Utricularia* spp., Lentibulariaceae) exhibits the most technically advanced trap, with highly effective architectural and operational features (Fig. 27). The submerged traps appear to be a modification of a foliar organ, although it sprouts on all parts of the shoot. They are quite small (mostly ranging between 0.2 mm and 6 mm, with the largest known at 10 mm). To its prey, the trap probably appears to be a small crustacean. But the most approximate analogy to describe the trap is to compare it to the wind instrument known as the bagpipe, except that the bladder of the bagpipe is under positive pressure, whereas the bladder of the trap is under negative hydrostatic pressure. A resulting additional difference is that inside air in the bagpipe is forced out under pressure through valves at the base of each pipe, which open outward when pressed by the musician; in the *Utricularia* trap, water from the external medium is forced under pressure into the bladder through a single, elaborate, technically advanced trapdoor that opens to the inside and is operated by a highly sensitive trigger mechanism when touched. Also, the pipes on the bagpipe are constructed and organized to produce music, but the appendages (antennae) surrounding the mouth of the trapdoor are designed to lure small prey and guide it toward the door. Some of the antennae are deceptively similar to filamentous algae, enticing prey to graze on them and approach the entrance to the trapdoor.

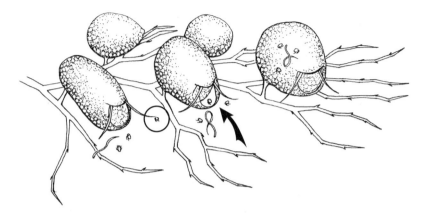

Fig. 27. Underwater traps of *Utricularia* (Lentibulariaceae) lie in wait for tiny crustacean prey. The trap is formed with negative internal hydrostatic pressure and a latched, hinged trapdoor. When triggered by the prey, the door springs open, and the prey is sucked into the trap.

In the primed trap of *Utricularia,* the negative hydrostatic pressure within the bladder cause its sides to become convex and bulge inward. When the trapdoor opens in response to activation of the trigger, water rushes into the bladder from the surrounding medium. Any animals caught in this deluge are sucked into the trap, which closes up extremely rapidly: ~0.01 seconds from triggering to closure. Relief from the negative pressure causes the sides of the bladder to become concave and bulge outward.

The elaborate and highly sophisticated structure and function of the trapdoor are the key elements in the operation of the trap, and they account for the extremely rapid response. The door opens inward and upward, supported by a thin, articulated hinge situated at the top of the mouth. Additional articulation is provided by another flexible hinge situated in the center of the door. When the door is shut, its base rests tightly against a sill made up of a specialized "pavement epithelium" of cells, organized in the form of cobblestones. The front edge of the bottom of the door is semicircular and is tightly pressed against a low, semicircular threshold of specialized cells, forming a tight seal. In addition, cuticle over the outer threshold cells separates and comes to rest pressed against the seal between the door and the threshold, forming an additional seal. The cells of the threshold inside the door also secrete mucilage, sealing the door from the inside.

In the primed condition, the door faces pressure from the outside, yet despite being designed to open inward it remains tightly shut until triggered open. This is made possible by the construction of the door itself, which consists of two layers of cells. The outer layer is composed of smaller cells that are virtually nonexpandable, while the inner layer consists of much larger cells that are highly flexible as a result of their infolding walls, which allow large-scale lateral expansion when the door is shut and contraction when it opens, like a bellows. Small changes in turgor pressure can cause large-scale changes in their volume.

As long as the trap is primed, excess turgor of the cells of the inner layer causes the door to press outward, sealing it against the frame, the pavement epithelium of the sill, the threshold, and the velum. However, at the lower edge of the door the cells of the inner layer are smaller and more compact, making it stiffer. Cells of the inner layer are arranged in concentric semicircles, centering at the bottom of the door. Cells in both layers of the wall of the bladder are green with chloroplasts, but the door itself is translucent, allowing a spot of light to illuminate the interior of the bladder. This may serve as an attractant for prey. The internal walls of the bladder carry numerous quadrifid glands that produce digestive enzymes when prey is present and then absorb the products of digestion.

A number of specialized, stiff bristles, situated immediately below the central hinge, extend outward from the door among the numerous antennae. These are trigger bristles that are excited by mechanical perturbation, causing the trapdoor to open inward. A mechanistic approach describes the trigger hairs as levers that are forced downward by the touch of an insect, causing the bottom of the trapdoor to move inward, opening a gap for the pressurized inrush of water into the bladder that forces the door to open to its widest extent. This explanation does not take into account the loss of turgor from the cells of the inner layer that accompanies the opening of the door.

An alternative electrophysiological approach describes the trigger bristle as a source of an electric signal that is transmitted to all the cells on the interior of the door causing their virtually instantaneous loss of turgor; the resulting imbalance forces the door to collapse inward. This approach is supported by existence of species of *Utricularia* that lack trigger hairs but have stalked glands instead. However, the transduction chain may not necessarily involve electric signaling. Pressure inside the bladder and outside are equalized rapidly, allowing the trapdoor to swing back passively and take up its position against the doorframe. The internal layer of the door now takes up water and rebuilds its turgor pressure concomitantly with the evacuation of water from the bladder and reduction of its hydrostatic pressure.

When the trapdoor opens, it does so by a sinuous undulation that is propagated upward from the middle of its lower edge. Water rushes in extremely rapidly and forcibly, increasing the volume inside the bladder by ~40 percent. As soon as the pressure differential (initially amounting to ~170 Pa) is equalized, or slightly in advance, the trapdoor slams shut. A single animal can activate the trapdoor, acting as a Judas goat by causing the forceful rush of water into the bladder that drags all innocent bystanders in the vicinity of the mouth to their fate inside the bladder. The entire process, from triggering to resealing, is extremely rapid, lasting about 30 milliseconds.

A number of processes start when the trap is closed. One is digestion of the prey by enzymes produced by the numerous quadrifid glands that line the entire internal surface of the bladder. However, the most essential process after closure of the trap is resetting to the primed state, with a negative hydrostatic pressure within the bladder. This means that about 30 percent of the water inside the resealed bladder is removed to the outside, against the increasingly steep gradient in hydrostatic pressure potential across the thin walls of the bladder. This is a two-step process, requiring a primary absorption of water by internal tissues followed by extrusion of the absorbed water to the outside.

This is a remarkable feat, the creation of a negative pressure inside the plant trap. Apparently the process takes place in the tissues adjacent to the bottom of the trapdoor, as the extruded water has been shown to accumulate around this location outside the bottom of the door. The volume of the extruded water corresponds approximately to the volume of water lost from inside the bladder. Obviously, water cannot pass out of the tightly sealed trapdoor. Responsibility for the primary step in this mass transport of water is allocated to the bifid glands that form a continuous carpet over the surface of the pavement epithelium just inside the bottom of the door. As a matter of fact, almost every epidermal cell of the pavement epithelium carries one such bifid gland.

Various explanations have been offered to account for the possible function of these glands. These must take into account the experimental fact that the outflow of water is not affected by lowering the water potential outside the bladder; rather, it is affected by similar changes in water potential inside the bladder. Clearly, membrane permeability controls the transport of water from inside the bladder into the inner layer of active cells. In contrast, any membranes separating the corresponding cells facing the external medium are fully permeable to solutes. Extraction of water from the lumen of the bladder into the internal layer (bifid glands) is accomplished by active transport of osmotically active solutes (primarily chlorine accompanied by sodium), energized by respiration, or/and photosynthesis. Water potential in the cells of this layer decreases, causing entry of water from the lumen and a consequent increase in their hydrostatic pressure. The increase in pressure moves water, but not the solutes, into the corresponding cells of glands in the external layer of the trap wall. Ions accumulating in these cells create a gradient in concentration toward the lumen of the bladder and are recycled by passive diffusion. The concentration of ions within the lumen of the bladder is not reduced during resetting of the trap, but actually increases. This supports the concept of recycling of the ions between the glands in the inner layer of the wall and the lumen. The increase in ion concentration could therefore result from the reduction in the volume of water as the bladder contracts during resetting. In the absence of a permeability barrier to the outside, bulk transport of water takes place when the hydrostatic pressure in these cells exceeds that in the external medium. This is the most acceptable explanation to date, but it is by no means the only one.

CHAPTER EIGHT

Liberation Movements: Leaving the Nest

PLANTS, like all other biological beings, are programmed for sexual reproduction and generation of offspring. It could be argued that plants excel at this hegemonic imperative if one considers the success of certain plants as agricultural species—corn (*Zea mays,* Poaceae) is a prime example of a plant now growing all over the world, over vast areas. Wild, native plants display a variety of developmental programs aimed toward producing offspring at the optimal season for their successful future establishment. Plants vary widely in the number of offspring they make and the way the offspring are packaged and dispersed. We humans and a host of other animals are both pollinators and dispersers, playing important roles in plant reproduction. Flowers and other reproductive structures exhibit a stunning creativity, adaptability, trickery, and attractiveness for this purpose. Ultimately, the offspring, either spores or seeds, are shed, ejected, carried, or wafted away in the wind to new settings.

Formation of Spores and Seeds for Dispersal

All terrestrial plants start their sexual generation as unicellular spores that develop into male, female, or bisexual gametophytes. Spores are produced within sporangia, protected by a wall that consists of a single layer or multiple layers of cells. Sporangia are produced on specialized leaves, the sporophylls. A few ferns and all seed plants (angiosperms and gymnosperms) are heterosporous; their microsporangia produce numerous small male microspores while their megasporangia usually produce a single, large female megaspore/macrospore (rarely four) that contains stored food. Megaspores develop into the female megagametophyte without leaving the megasporangium and its remaining (diploid) sporogenous cells that serve as additional protection and source of nutrition. In this condition, the megasporangium is called nucellus and is provided with additional physical protection by two concentric tissues that grow up and all around it from its base (integuments), leaving a narrow pore at its tip (micropyle). For flowering plants (angiosperms), the latter serves as port of entry for the pollen tube that brings the male gamete (sperm) into contact with the female gamete (egg cell) for fertilization. The nucellus and its integuments are the ovule. The fertilized egg cell develops into the (diploid) embryo, the ovule becomes a seed, and its integuments become the seed coat.

In seed plants, microspores develop into pollen grains, and the microsporangia are pollen sacs. In the gymnosperms, to which conifers belong, the microsporophylls are organized in a tight spiral to form male cones; the ovule lies exposed on top of the megasporophyll, which is also organized in tightly packed, thick, lignified bracts, arranged in cones that provide very effective protection during ovule and seed development and maturation. Hyponastic growth of the sporophylls (male and female) seals them to each other. These spread apart (by epinastic growth) after the ovules are fully mature, allowing access to pollen; they then move tightly together again (hyponasty), to spread open once more (epinasty) when the seeds are ripe, allowing them to fall freely.

In the angiosperms, the megasporophylls transform into a carpel, either by fusing opposite margins to each other or by fusing with the margins of their neighbors on either side. In either case, the seeds are enclosed within a special container—the ovary—which eventually matures into the fruit (a process closely associated with seed development). An ovary may consist of one to several compartments, each formed by a single carpel with its opposite margins fused, or by two or more carpels with neighboring margins fused, or by a number of separate carpels, each with its opposite margins fused. Organs that enclose the ovary (floral organs and even external bracts) may also undergo transformations and combine with the fruit to form more complex dispersal units. Ovaries are characteristically suspended along the placenta at the fused margins.

Once formed, spores and seeds have no further need for nutrients, water, or growth substances from their parents. Supply routes between them and their parents are inactivated. To perform their function in life, they have to be able to move freely and as far as possible from their parent. They are now primed to become liberated, to disperse into their environment, and to start a new life as independent plants. Seeds formed by fleshy fruits (common in Cucurbitaceae such as melons and squash, in Solanacea such as tomatoes, in Rosaceae such as apples, pears, and strawberries, and in Vitaceae such as grapes) or diasporas (seeds surrounded by other tissues) are usually liberated by senescence and disintegration of their succulent tissues; frequently they are aided by harvesting animals. In most dry fruits, liberation starts with differentiation of an abscission layer across the stalk on which the seed is suspended. Formation and performance of this layer are very similar to those of the layer formed across petioles of leaves in deciduous plants when senescent leaves become detached and fall. Some single-seeded fruits remain closed, and the fruit (or dispersal unit) is detached entire. In nuts, acorns, and the grains of grasses (caryopsis), the fruit coat is fused with the seed coat, sometimes remaining enclosed by its specialized floral bracts. Abscission is the final step, as the embryo germinates from within the fruit. Dispersal units consisting of a number of flowers (florets), each with a single-seeded fruit, may also fall into this category: for instance, *Rumex roseus* (Polygonaceae) and the spikelets of certain grasses such as barley (*Hordeum* and *Aegilops*, Poaceae).

Breaking Down the Walls

In sporangia (and anthers) and dry fruits, the seeds are liberated posthumously, as the fruit coats (pericarp) senesce and die after maturation. The coats dry out and shrink. Shrinkage is the result of evaporative loss of hydration and capillary water from the cellulose network of the coat. The increasing strain (tension) eventually causes the coats to split open, liberating the seeds. In some fruits (for instance, the pomegranate *Punica granatum*, Punicaceae), the coats eventually split open haphazardly at unpredictable locations by isotropic shrinkage against the succulent seeds packed tightly within it.

Forcing Open the Gates

In most dry fruits, dehiscence is well orchestrated far in advance and takes place along sutures located in predestined locations and tissues. A ventral suture is located at the juncture of two adjoining margins of the same carpel or of neighboring carpels and is a favorite site for the fault line. For loculicidal dehiscence in most plants, this is the location of the placenta, where the ovules/seeds were attached. Another favorite location for the fault line is the dorsal suture along the plane of symmetry of the carpel, down the middle of its conducting tissue (septicidal dehiscence) such as the capsule of *Iris* (Iridaceae). Certain dry fruits exhibit both ventral and dorsal sutures, as in many Fabaceae such as the pea (*Pisum*).

A suture is split by means of a motor located in adjacent tissues of the pericarp. The motor acts by generating a force of contraction (stress) transverse to the direction of the suture. Pericarp tissues adjoining the suture exhibit long, narrow cells with their axis parallel to the suture, and cell walls with their cellulose microfibrils parallel to each other and to the suture. As the cells dry after maturation, the dehydrating microfibrils and increasing surface tension in their microcapillaries cause the cells to contract transverse to the axis of the microfibrils, creating progressively increasing strain. The suture splits when the strain exceeds the adhesion along a fault line. The increasing tension pulls apart the coat on the opposite sides of the suture, splitting it open. An additional function of the motor is the result of presence in the pericarp of an internal and external tissue, distinguished by the different structure. As a result of these differences, contraction of the external layer is antagonistic to that of its subtending, internal layer, in addition to drying out more rapidly. The coat unfolds outward, increasing exposure of the enclosed seeds and facilitating their liberation.

The suture splits lengthwise by means of a fault line that bisects it and has the lowest mechanical resistance, because the adhesive properties of the middle lamella that holds together the cell walls are weakest. A fruit with more than one suture splits open into valves. Use of sutures for dehiscence is not universal. Certain fruits may dehisce along fault lines that have no relation to the carpels. For instance, in the poppy (*Papaver*, Papaveraceae) and the snapdragon (*Antirrhinum*, Scrophulariaceae), dehiscence takes place by means of flaps near the top of the capsule. In the bouncingbet (*Saponaria officinalis*, Caryophyllaceae), the capsule dehisces by means of teeth-like flaps at its top; the outer walls of the epidermis are thickened inward in transverse folds, and its contraction in opposition to the subepidermal layer causes the teeth to split apart and curve out, exposing the seeds. In the scarlet pimpernel (*Anagallis arvensis*, Primulaceae), the capsule dehisces by means of a cap that detaches from the top of the capsule. In the mustard (*Sinapis arvensis*, Brassicaceae), the base of the fruit (siliqua) dehisces by means of two valves along sutures joining the margins of two opposite carpels, but the upper part of the fruit remains and is dispersed indehiscent.

Moving Out

Cohesion-Mediated Seed Propulsion

BALLISTIC PROPULSION BY CATAPULT. In most dry dehiscent fruits, such as capsules of *Lilium* (Liliaceae) and *Iris* (Iridaceae), the seeds are simply liberated by progressive, basipetal dehiscence and disperse passively, either by gravity alone or with the aid of wind, rain or animals. However, in many plants with dry fruit, the seeds are not simply liberated to disperse passively but are actually boosted ballistically by explosive movements of the fruit coat at the time of dehiscence (Fig. 28). Examples include Acanthaceae, Balsaminaceae, and Euphorbiaceae; *Alstroemeria,* Alstroemeriaceae; common among Papilionaceae such as *Lotus;* and Fabaceae such as *Bauhinia, Wisteria,* gorse (*Ulex* spp.), and brooms (for example, *Cytisus scoparius, Genista fasselata,* and *Spartium junceum*).

Among the Euphorbiaceae, many species exhibit capsules that dehisce explosively. In the castorbean (*Ricinus communis*), the fleshy exocarp dries out and deteriorates, leaving a lignified endocarp that dehisces basipetally into three one-seeded valves with transverse fibers. As these become progressively dehydrated, they eventually reach a strain threshold

Fig. 28. Scotch broom (*Cytisus scoparius,* Fabaceae) forcefully ejects seeds from its pods, which are constructed to pop open when dry.

beyond which they recurve instantaneously, catapulting the seed (~3.5 m). Similar dispersal techniques occur in shrubs such as *Euphorbia marginata* and *Mercurialis perennis* (~4 m). In *Hevea brasiliensis,* as the large fruit dries out, the pericarp detaches first from the central columella. With the progress of dehydration, the increasing strain causes the pericarp to split longitudinally into three valves along the ventral suture from its top down. Additional strain eventually causes each valve to split longitudinally along its dorsal suture. This movement is instantaneous: each half valve twists a bit around its axis, causing the seeds to be catapulted out (~36 m). Large trees catapult their seeds to greater distances: *Hura* holds the record for distance (~45 m).

Plants in the entire family Acanthaceae are distinguished by explosive dehiscence of the two-valved capsule. At maturity, each capsule is enclosed between two lignified sepals and located in the axil of a lignified bract. These clasp the capsule tightly when dry, opening up after rainfall to expose the capsule. Each valve is under increasing strain as it dehydrates after maturation due to structural differences in its exterior and interior layers. Eventually the sutures on either side split open explosively along their fault line. In *Blepharis ciliaris* and *Ruellia prostrata* the valves are held together at their tip by a plug of parenchyma; this allows them to split open along a fault line by imbibition of the middle lamella after rainfall. The seeds are catapulted out their capsule at time of dehiscence, but not by means of the contortions of the valves. Each seed is attached at its base to the placenta by means of its stalk (funiculus = retinaculum), which is lignified and compressed as a spring by the seed above it at maturity. The strain in this organ is relieved instantaneously by the explosive separation of the valves, enabling it to propel the seed as a projectile to a significant distance (~9 m in *Acanthus mollis* and *A. longifolius*). A similar explosive mechanism occurs in *Alstroemeria* (Amaryllidaceae); however, the seeds are catapulted out by release of strain in their subtending, lignified placenta. The achene of *Polygonum virginianum* (Polygonaceae) is attached to its stalk by a separation layer, consisting of a peripheral, inelastic sheath of vascular tissue surrounding a central core of parenchyma (pith). Pith cells continue to expand within the vascular cylinder, compressing their upper layers against the transverse partition of the separation layer. Mechanical perturbation of the style above the achene results in disruption of the separation layer, enabling its compressed parenchyma to expand explosively, catapulting the seeds.

Structural differences in parallel tissues of the fruit coat are employed in the motors. In pods of the subfamily Papilionaceae, such as *Vicia* and *Lupinus,* the elongated cells of the exocarp and endocarp are not laid down with their long axis parallel to the suture but rather are at an acute angle (£45°) with it, precisely opposite in the exocarp and endocarp. As a result, as these tissues dehydrate, each of them contracts along transverse planes that are at an angle to each other, creating progressively increasing strains that act at an angle to each other and to the suture. The fault line is exposed to the combined increasing strain until adhesion is overcome at its weakest points. An instantaneous surge of tension is generated uniformly along the entire length of the suture, allowing the strain to transform into potential kinetic energy. The opposite (symmetrical) parts spring apart explosively. At the same time, tension is also released in the external and its underlying tissue of the valves, but this takes place at an angle to each other. As a result, the coat springs back explosively into a spiral, its preprogrammed, relaxed configuration. This choreography causes the seeds to be catapulted out forcibly for considerable distances away from their parent. A similar mechanism exists in *Wisteria sinensis* and *Lupinus angustifolius,* where the exocarp is composed of two layers with similar fiber orientation, but their cellulose microfibrils are in opposite angles of orientation. In contrast, in *Astragalus fruticosus, A. hamosus,* and *Hedysarum pallens* the endocarp exhibits two parallel layers with divergent angles of orientation of their fibers.

Viola canina (but not in other species of *Viola,* Violaceae) and witch-hazel (*Hamamelis,* Hamamelidacea) provide another variant of triggered catapults. The capsule dehisces first, and seeds are catapulted out only after the (three) valves dehydrate to reach their convulsion threshold. An analogous process has been recorded in dispersal units (that include the perianth or inflorescence) of *Dorstenia* and *Sloetia* (Moraceae), or dwarf mistletoe (*Arceuthobium,* Viscaceae), where the catapult motor acts by dehydration and shrinkage of floral tissues that enclose the fruit coat. In *Phlox drummondii* (Polemoniaceae), the capsule is held together by its surrounding calyx. As it ripens and dries out, the calyx contracts abruptly, pushing the capsule out and allowing the valves to spring apart explosively, catapulting the seeds (to ~1 m). The catapult motor in *Dictamnus fraxinella* (Rutaceae) employs a different technique. The five carpels are spread radially and dehisce along their upper (free) edge as they ripen, exposing a stiff free endocarp, detached from the exocarp. Ripening progresses until tension in the endocarp passes a threshold value, at which point each endocarp twists abruptly at its base, separating into two halves and catapulting the seeds out (~1–2 m).

CENTRIFUGAL PROPULSION.   The fruit of *Geranium* (Geraniaceae) consists of five separate, single-seeded mericarps that hug the base of their common receptacle, resembling a stork's bill. The latter extends upward as a thin beak, to which the separate styles of the five individual carpels adhere. The distal parts of the mericarp are fused together, a continuation of the receptacle. The styles exhibit an external sector with elongated, thick-walled cells (and cellulose microfibrils) transverse to its axis, and an internal sector with cells parallel to its axis. As the mature fruit dries out, the external sector contracts along the axis of the style, against resistance by the internal sector. The strain that is produced eventually causes abrupt splitting along fault lines separating the mericarp from the receptacle, proceeding acropetally from the base of the receptacle to the fused styles. The liberated styles coil upward, remaining fused at their tips to the top of the receptacle. This violent action produces centrifugal forces by which the seed, exposed at the base of the (open) mericarp, is ejected and propelled to some distance (~2–3 m).

The mechanism of dispersal in the related *Erodium* (Geraniaceae) is based on a similar structure of the fruit, except that the structure of the style is different. The proximal portion (~⅔) of the dry style consists of two parallel tissues. An external layer with thick-walled, elongated fibers (and cellulose microfibrils) is oriented at an acute angle to the axis of the stork's bill receptacle. An internal layer consists of thin-walled, shorter cells, with their long axis along the axis of the receptacle. Thus, upon drying, the differential contraction detaches the style from the receptacle and forces its proximal part to contract in a tight spiral, leaving its distal part straight. The entire mericarp is detached intact and is dispersed passively.

Alyssum spp. (Brassicaceae) provides an example of a centrifugal catapult. The motor is located in the flower stalk (peduncle). The trigger is primed by dehydration after ripening, when the peduncle curves sharply down by differential expansion, greater in its adaxial (upper) than in its abaxial (lower) tissues ones. The former is characterized by a layer of thick-walled fibers, oriented transverse to the axis of the peduncle, while the latter is composed of thin-walled cells acting as anchor. A similar situation occurs in *Salvia horminum, S. lyrata,* and *Teucrium euganeum* (Lamiaceae). Curvature of the peduncle takes place by contraction of the abaxial sector of the pedicel, resulting from transverse orientation of the fibers, in contrast to the longitudinal orientation of the fibers in the adaxial anchoring tissue. Curvature is accompanied by radial contraction of the calyx tube, resulting from isotropic contraction of its inner epidermis, and its teeth bend inward, trapping the fruit. When wet by rainfall, the curvature of the peduncle is reversed: the corolla tube widens, its teeth straighten, and the trigger is released abruptly, resulting in forcible ejection of the fruit by the centrifugal force generated by the catapult.

Turgor-Mediated Seed Propulsion

In all the fruits so far described, the forces acting on the fruit coat are those produced by adhesion and cohesion. A number of fleshy fruits (such as *Impatiens parviflora,* Balsaminaceae; *Ecballium elaterium,* Cucurbitaceae; and *Cardamine impatiens,* Brassicaceae) remain live and fully turgid until they dehisce explosively. Turgor pressure powers the motor. *Biophytum* and *Oxalis acetosella* (Oxalidaceae) employ such explosive dehiscence to propel their seeds as far as 2 m. The parasitic *Lathraea clandestina* (Orobanchaceae) catapults its seeds to distances of 12–20 m.

JET PROPULSION. *Ecballium* (Cucurbitaceae) has a fleshy, green, ellipsoidal fruit borne on a stalk that is inserted into the top of the fruit and is curved downward (positively gravitropic). Underlying the external photosynthetic tissue are several layers of ellipsoidal cells with thick walls, rich in pectic substances. The internal tissue, in which the numerous seeds are embedded, is composed of very large, thin-walled cells whose cytoplasm continuously loads the central vacuole with a glucoside—elanterinidin. The continuous accumulation progressively increases the turgor pressure in the entire tissue. The mounting pressure is transmitted outward by the surrounding, thick-walled cells to the fault line around the insertion of the stalk into the fruit, which eventually ruptures instantaneously; the seeds are ejected explosively as a jet stream (to distances frequently exceeding 2 m), accompanied by the slimy content of their surrounding cells that exploded simultaneously. The slime enables the seeds to attach to any solid surface.

In contrast, fruit of the related *Cyclanthera explodens* is not fleshy, but eventually it splits up into three valves that curve outward simultaneously and explosively, thereby catapulting the seeds out. A somewhat different means of jet propulsion has been attributed to *Oxalis acetosella* (Oxalidaceae). Dehiscence of the pod is independent of the subsequent liberation of the seeds. The latter develop within massive mucilaginous coats that combine to fill up the entire pod. The mucilage dehydrates and separates from the seed proper, eventually squeezing the naked seeds explosively out of the splits in the coat.

TURGOR-MEDIATED CATAPULTS. In *Impatiens* (Balsaminaceae), the fruit is an elongated, cylindrical capsule with a fleshy pericarp composed of four or five carpels (valves in the fruit) joined margin to margin. The subepidermal tissue of the pericarp consists of live, tightly packed, large, thin-walled parenchymatous cells that are elongated in anticlinal orientation. The underlying tissue consists of thick-walled, narrow cells that are elongated in periclinal orientation along the fruit axis (at right angles to the former), parallel to the sutures. Turgor pressure builds up during maturation, increasing the strain on the fault line in the sutures. During maturation, turgor builds up in cells of the subepidermal tissue cells. They expand, exerting increasing strain transverse to the fault lines in the sutures along the underlying tissue. Eventually the strain splits the sutures along their fault lines. The separation is instantaneous, releasing the cells from constraints on expansion (elongation ~32%), which results in explosive inward curvature of the valves. The attached seeds are catapulted to long distances.

Spores

In all heterosporic plants (some ferns and all seed plants—Spermatophyta, including the gymnosperms and angiosperms), haploid microspores are formed (by meiosis) within microsporangia. In seed plants, the microspores are pollen grains; the microsporangia are paired pollen sacs borne on microsporophylls. In gymnosperms, such as conifers, the microsporophylls are arranged in a tight spiral, forming a male cone. In angiosperms, the microsporophyll is part of the flower. They characteristically consist of a long, slender, elastic filament topped by an anther consisting of two paired pollen sacs.

   Pollen (microspores) can only perform their mission after liberation from the sporangium or pollen sac (microsporangia) in which they developed after meiosis. Spores are much smaller and lighter than seeds by several orders of magnitude. Their surface/mass ratio is very high, enhancing the frictional effect of air movement. Spores must be liberated to enable their free movement to their eventual destination. After their liberation, they are carried over long distances by the lightest air currents, which is the preferred means for their dispersal.

   Spores of lower plants (mosses and liverworts) germinate on virtually any moist substrate. This is readily available in their natural habitat, and most of the spores produced manage to germinate and grow into gametophythes. Pollen grains can only perform their destiny by germinating in close proximity to their target—the megaspore. The vast majority lack any mechanism to guide their movement through the environment to their destination. They never make it and expire unfulfilled. This tremendous wastage is countered by mass production in vast numbers. Exceptions are plants that have established a symbiotic relationship with certain nectar-collecting animals that act as vectors and transport their pollen directly to its proper destination.

COHESION-MEDIATED SPORE PROPULSION.   The walls of sporangia also have well-defined fault lines for breaking open in response to their contraction by dehydration. Isospores of mosses, liverworts, and ferns (and their relatives) as well as microspores of all other plants (with a few exceptions, such as orchids) are released into the atmosphere by rupture of the walls of their sporangia, caused by shrinkage as they dry. Such rupture takes place along predetermined fault lines, where the adhesive holding the adjacent cell walls is inherently weak. Sporangia (capsules) of liverworts (such as *Marchantia,* Marchantiaceae; and *Pellia,* Pelliaceae) contain two types of cells. The sporogenous tissue forms haploid spores by meiosis and elaters that remain diploid. The drying capsule characteristically (but not universally) splits open by longitudinal slits.

Elaters act as motors by which the liberated spores are expelled violently as a result of their extensive contortions during dehydration. These are slender, elongated dead cells with walls reinforced by spiral thickenings. As the elaters dry, they contract lengthwise; their spiral thickenings cause them to twist and contort extensively in the process. As hydrostatic pressure in the lumen becomes increasingly more negative, the increasing tension causes water to be withdrawn from the microcapillaries in their walls. Eventually, the largest microcapillaries evacuate completely, allowing air to enter explosively (embolism), at which point the elater relaxes abruptly to its straight configuration. Spores within reach of the elater are propelled violently into the surrounding air. The perturbation created by the violent movement of the elater causes perturbation of its neighbors, who respond by violent relaxation of their own. Elaters of *Cephalozia* (Cephaloziaceae) are anchored to the internal walls of the capsule until they are released explosively upon their abrupt relaxation during dehydration.

In most mosses (such as *Funaria,* Funariaceae), the drying capsule dehisces by separation of a cap (operculum) at its top, revealing a ring (annulus) of tooth-like or other appendages (peristome) over the open capsule. The appendages consist of two parallel tissues: an external layer, with thick-walled fibers in parallel, transverse to the axis; and an internal layer with thin-walled cells in parallel to the axis. Dehydration contracts the external layer along the axis, and hydration causes its expansion. This causes the appendages to curve out when dry, exposing the spores, and to curve back when wet to close up the capsule and prevent spore dispersal.

In contrast, the capsule in *Sphagnum* (Sphagnaceae) has no peristome and remains sealed. As the capsule dries out, its contents are under progressively subatmospheric pressure (as great as -50 bar or -5 MPa). Eventually,

a threshold value is reached beyond which outside air enters the contracted capsule, causing it to inflate explosively, accompanied by a mass
discharge of spores (Fig. 29).

Sporangia of isosporous ferns, such as maidenhair (*Adiantum capillus-
veneris,* Pteridaceae), woodfern (*Dryopteris,* Dryopteridaceae), or brake
fern (*Pteris* spp., Pteridaceae) are ovoid. The wall of the oval sporangium
consists of a single layer of flat cells attached to the sporophyll by means
of a short stalk. A single row of isodiametric epidermal cells is differentiated along the crest of the sporangium, forming the annulus. Approximately three quarters of the way around the sporangium, a transverse
fault line intersects the annulus. Upon dehydration, the sporangium splits
open along the fault line and folds open progressively. Beyond a certain
threshold, the direction of this movement is reversed abruptly, catapulting the spores violently. The annulus acts as the motor for the movements, by means of specialized structural features of its cell walls.

Cells of the annulus are distinguished by U-shaped thickenings along
their adjoining walls and across their bottom. As water evaporates from
these cells by dehydration, they contract anisotropically, primarily along

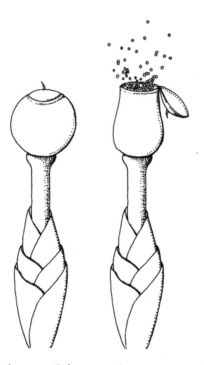

Fig. 29. Spores of *Sphagnum* (Sphagnaceae) moss are propelled out of a
capsule when the cap dries and springs open.

their thin upper walls, narrowing the gap between their thick lateral walls. As dehydration proceeds, negative hydrostatic pressure develops in cells, and the water is under increasing tension. When the strain that develops along the annulus exceeds the mechanical resistance of the fault line, the latter pulls apart. The annulus is now free to change progressively from convex to concave, leaving the spores exposed in the coat. The increasingly negative hydrostatic pressure inside the lumen of the cells eventually exceeds the surface tension in the (widest) microcapillaries of their walls, which then enables air to rush in under high pressure, causing embolism. The cell expands explosively, thereby jolting its neighbors to embolize sympathetically. The entire annulus, acting as a unit, becomes a catapult, scattering the spores far and wide into the air as it reverts to the relaxed convex configuration.

A somewhat similar mechanism acts in dispersal of pollen from anthers of most angiosperms. The motor is located in a specialized subepidermal layer, one cell in thickness, that surrounds the pollen sacs. Cells of this endothecium have thick walls in their base that extend upward into parallel, rib-like extensions and leave the external walls thin. As the anther dehydrates, cells of the endothecium contract anisotropically. Contraction is limited to their (upper) thin walls on their exterior, pulling the ribs closer together. Lateral strain develops over the surface of the pollen sac, which eventually causes its coat to split along the fault line, where it borders on its twin pollen sac. The coats of the two sacs fold back in opposite directions, exposing and liberating the pollen. Further dehydration causes embolism in the endothecium, which then springs back elastically, catapulting and scattering the pollen grains.

In exceptional cases, the coat of the anther is ruptured by means of pores (heath family, Ericaceae) or valves (barberry family, Berberidaceae). In the vast majority of higher plants, the coat of the anther is ruptured by longitudinal slits.

TURGOR-MEDIATED SPORE PROPULSION. In some species of insect-pollinated plants, pollen is targeted to the body of the pollinator, using movements by the anther. In the Papilionaceae subfamily, the typical corolla exhibits bilateral (zygomorphic) symmetry and consists of five petals. The margins of the uppermost (symmetrical) pair (the keel) adhere together, forming a sealed container in which the filaments of the stamens, joined side by side to form a hollow, elastic tube, are imprisoned. The margins act as a fault line, and they separate abruptly when an insect perturbs the flower, enabling the filament tube to relax and curve upward abruptly and forcibly. The architecture of the flower obliges the insect to land on the lowest (single and largest) petal—the banner—in a position where its body will intercept the liberated anthers and be showered with pollen.

A somewhat similar but much more spectacular mechanism is found in the triggerplant *Stylidium* (Stylidiaceae), characterized by a unique floral column in which the stamen and style are fused, and typically reside beneath the plane of the bilaterally symmetrical (zygomorphic). The basal parts of the four petals are joined by their margins to form the "throat" of the flower. The floral column protrudes from the throat and rests below the plane of the flower petals. The column exhibits two sharp, opposite bends in opposite directions along its plane of symmetry. The passive part of the motor consists of thin-walled, parenchymatous cells, while the core of the motor part at the bend consists of unlignified, thick-walled collenchymatous cells that lack intercellular spaces. The floral column is thigmonastic, acting as a trigger: it responds to the change in pressure when a pollinating insect lands on the flower by an action potential and immediate change in its turgor pressure. This causes the lower (proximal) bend to relax to become straight, sending the column flying forcibly toward the insect, covering it with pollen. The response to touch is extremely rapid, covering ~465° in 17.7 milliseconds (angular velocity ~26°/ms). After firing, the trigger reverts to its original bent configuration within a few minutes to a half hour. The turgor-mediated motor of the trigger is able to fire many times before it no longer responds to stimuli.

Self-Propelled (Automotive) Seeds

Diaspores of a number of different species are equipped with specialized organs that act (posthumously) as motors that propel the seeds in them over considerable distances over the soil surface and inside the soil in response to diurnal fluctuations in atmospheric humidity. The amplitude of these fluctuations is greatest at the soil surface due to radiative heating during the day and radiative cooling at night. These motors are powered by the nearly inexhaustible forces acting between water molecules utilizing the properties of cohesion (water molecules acting together) to perform movements.

Essentially, the motors for these automotive movements have striking similarities. They are located in long, slender appendages with differential structural features in tissues along opposite flanks, one of which acts as the operator and its opposite as the anchor. Structurally, the operator consists largely of parallel fibers aligned transversely to the axis of the motor. An increase or decrease in its hydration results in extensive expansion or contraction along that axis. Its opposite, anchoring tissue is structured to retain its dimensions unchanged, while allowing its neighboring motor tissue to undergo major changes along the axis, leading to curvature (frequently extensive). The diaspores are also equipped with a guidance system that makes their movements unilateral and forward.

In several species of Asteraceae (*Centaurea*), Fabaceae (*Trifolium*), and Dipsacaceae (*Scabiosa, Knautia*), the upper (distal) part of the seed or of the diaspore is equipped with stiff appendages (such as hairs, barbs, or crown) that face outward and toward the rear. These appendages are composed of two parallel, structurally opposite tissues. One of these (usually the external) is composed of cells organized parallel to each other, with their long axis and constituent cellulose microfibrils transverse to the axis of the appendage. This is the motor tissue. It expands or contracts along its axis in response to hydration and dehydration, respectively. The opposite, parallel tissue has a very limited capacity for expanding or contracting and thus acts as an anchor for its motorized neighbor. The opposite ends of the diaspore rest on the soil surface. The diurnal fluctuations in atmospheric humidity cause these appendages to move inward and outward at least once in each cycle, alternately increasing and decreasing the separation between the forward-oriented tip and the rearward-facing appendages. The rearward orientation of the appendages results in unilateral forward movement of the diaspore, particularly where its tip is also equipped with hairs or barbs that also face the rear and effectively prevent any movement in this direction.

Seeds of *Tamarix* (Tamaricaceae) are equipped with appendages consisting of a tuft of hairs. The tissue along the outward-facing sector of the (dry) hair exhibits tightly packed transverse folds of the thick outer walls of the epidermis over its basal portion. The fruit of *Anemone pulsatilla* (Ranunculaceae), a nutlet, is similarly equipped with hairs. A massive motor tissue occupies the outward-facing sector at the base of these hairs.

Diaspores of *Erodium* (Geraniaceae) come to rest on the soil surface on their opposite ends: their heavier tip, containing the seed and the straight, stiff, distal part of their style. The motor is located along the tightly coiled proximal portion of the long style. The transition from the tightly coiled, proximal portion and its straight, stiff distal part is almost at right angles, resembling an elbow with a proximal muscular motor controlling the movement of the stiff distal arm. The diurnal fluctuations in atmospheric humidity cause the motor to undergo alternate cycles of extreme coiling and uncoiling, which propels the base of the mericarp, with its seed, in repeated helical movements forward. The base is equipped with a cluster of stiff bristles facing the rear that ensure unilateral movement forward.

The diaspore of *Avena sterilis* (Poaceae) consists of paired spikelets enclosed between two glumes (green bracts). Each is equipped with a long, stiff bristle—the awn. The latter dries up during maturation into a tightly coiled, proximal portion, terminated by a straight distal portion. These bear a close resemblance to the beak of *Erodium* and exhibit an almost identical spiraling/straightening response to the diurnal fluctuations in humidity. They are also similarly equipped with a cluster of stiff bristles at their base, all facing the rear. The major difference lies in the fact that the *Avena* diaspore is equipped with two differentially oriented motors, rather than a single one. As a consequence, it tumbles during its forward propulsion.

Automotive diaspores move along the soil surface. However, when they eventually encounter an opening, or fracture in the surface, their forward tip inserts itself, whereupon further diurnal movements act to propel the diaspore deeper into the soil, over progressively shorter distances, as a result of increasing mechanical resistance by the soil matrix.

Blepharis persica (Acanthaceae) is an annual plant that inhabits hostile desert environments. Seeds of this plant as well as others like *Hura* (Euphorbiaceae) are propelled explosively when the trigger mechanism that locks the valves of its capsule until wet by rainfall (Fig. 30). As soon as the seeds land on wet soil, they germinate and become established within a few hours. They achieve their rapid establishment by a spectacular

Fig. 30. *Hura* (Euphorbiaceae) seed pods explode on drying, liberating seeds to fly some distance from their parent before starting a new life.

response to liquid water. Their coats are equipped with a dense cover of highly specialized, unicellular, long epidermal hairs, which in the dry state are all flattened back. The hairs are progressively longer from the tip (radicle end) of the seed to its opposite end. The walls of these hairs are composed of a highly hygroscopic matrix that can imbibe and swell repeatedly. At the same time, the walls are reinforced over their entire length by spiral thickenings that do not swell. The distal tip of each hair is split into separate short extensions. As soon as the seed is in contact with liquid water, the hairs start swelling, straightening out in the process and thereby raising the seed above the soil. The seed is tilted with its radicle tip resting on the soil surface at a 45° angle, the result of the progressive tip to base increase in length of the hairs. As a result, the emerging radicle is poised at a most favorable angle for insertion into the soil.

Conclusion

Elizabeth Van Volkenburgh

I F PLANT MOVEMENTS were accompanied by sound, a cacophony would greet us as we step into the garden, the field, or the forest. Such is the effect of discovering the pervasiveness and diversity of movement activities going on among plants. The more obvious, above-ground movements of sun tracking and phototropism are familiar to many. Surprises presented here include the burrowing, contractile, and sensory behavior of underground plant parts as well as the mysterious determination of branch angles constructing canopies and root masses. Going be-yond a description of plant morphology and the sudden movements of sensitive plants or traps, Dov Koller has incorporated growth physiology into the discussion of plant movements, thus reinforcing the notion of the restless plant, one that is constantly moving, seeking, sensing, and respond-ing to its environment.

The mechanisms driving plant movements described here are based on a physiological understanding of plant cell function. Fundamental are two characteristic features of plant cells: (1) a rigid cellulosic cell wall bounds all plant cells, and (2) living plant cells generate positive pressure by osmosis because they are hydraulic structures. Although molecular motors associated with movement in animal cells are also present in plants, including actin and myosin, the macroscopic movements that catch our attention are driven by changes in cell turgor (pressure) due to water uptake or loss, and by hygroscopic or other changes in the orienta-tion or shape of the cell wall. Differential expression of these traits, in various tissues at different stages of development, has given rise over the course of evolution to the wide variety of movements we now recognize.

For example, compare the sensitive *Mimosa pudica* (Fabaceae) with a relative whose leaves do not move in response to touch, such as the silk tree *Mimosa (or Albizia) julibrissin.* What makes the difference? Has the sensitive plant harnessed a mode for creating and transmitting electrical signals from a touched leaflet to its pulvinus, but the silk tree's leaves have not? Or is it that, unlike its relative, the sensitive plant's pulvinus has incorporated a different sort of membrane protein that can respond to an incoming electrical signal? Answering this type of question will help us understand how the mechanistic parts have been selected and assembled to provide a diversity of movements among species. The wide range of taxonomic groups mentioned in this book is no doubt the tip of the iceberg for discovering movements in plants. Phylogenetic or comparative studies promise discovery of not only the physiological but also the molecular and genetic bases for movements in plants and their adaptive significance.

Plant movements can be used as case studies in plant behavior, providing a window into the "black box" of how signals, both external and internal, cause responses at the molecular, cellular, organ, and organismal levels. Take, for example, the sun-tracking leaf movements in *Lavatera* (Malvaceae), one of Dov Koller's best-investigated species. In these leaves, light is perceived by the leaflet blade (lamina), and movement is generated in the pulvinus at the base of the leaflet. Many questions remain unanswered. What photoreceptor is responsible for light perception? Upon excitation, what cellular response does the photoreceptor induce in the perceiving leaf cell(s)? How is this response transmitted from the illuminated cell to its neighbors, then through the veins to the pulvinus? Which cells in the pulvinus perceive the incoming signal? What does perception of the incoming signal accomplish there, and how does that cause the (relatively well-understood) osmotic response leading to pulvinus deformation and leaf movement?

From current literature, it is reasonable to hypothesize that the photoreceptor is phototropin, one of the blue-light photoreceptors. A number of cellular reactions occur in different cell types after excitation of this molecule, including interaction with 14–3–3 regulatory proteins, phosphorylation and activation of the plasma membrane proton pump, interaction with motor proteins leading to chloroplast relocation, and possibly control of ion fluxes through ion channels. In *Lavatera*, if the phototropin is located in parenchyma lining the vascular tissue, which of these processes is it influencing there—or is there perhaps something else going on? During phototropism of seedling stems, excitation of phototropin causes redistribution of polar auxin transport machinery. Is this how phototropin influences the pulvinus in *Lavatera*, by altering the flow of auxin from the blade to the pulvinus? This raises the question, Is the *Lavatera* lamina communicating with the pulvinus by an electrical or chemical (auxin) signal?

In contrast to *Lavatera*, bean leaf movement is driven by light sensed directly in the pulvinus cells. In this case, it is known that phototropin is localized in the cells that swell/shrink to move the pulvinus, and the photoreceptor interacts biochemically to regulate ion flux and osmoregulation. It has also been reported that auxin is involved within the motor cells when osmoregulatory changes drive leaf movement. Interestingly, in the soybean (*Glycine max*, Fabaceae) there is a range of genotypes differing in their ability to track the sun. Soy leaves orient at right angles to the sun in early morning and late afternoon, but during midday they orient parallel to the sun. Some lines of soybean carry out these movements more accurately than others. Evidence indicates that the accuracy of leaf movements leads to higher yields in soybean fields by minimizing potential damage to the photosynthetic apparatus during midday caused by high light intensity. What is the genetic basis for the range of accuracy in sun tracking among these soybean lines? Possibly the difference lies in localization of phototropin in membranes of pulvinal cells; alternatively, the osmoregulatory responsiveness of pulvinal cells to phototropin could vary due to slight changes in the activity of one of the intermediate signaling molecules (e.g., the 14–3–3 protein or auxin receptor). Discovery of the physiological mechanism at the base of this variation in behavior of soybean leaves not only would augment our understanding and improvement of yield in soybeans but will be immediately relevant to questions of how auxin regulates growth of stems and other tissues during plant development.

Apart from growth regulation and movements of the pulvinus and other tissues that are analogous to it, two other types of plant movements provide useful portals for figuring out how plants sense and respond to environmental inputs. Dov Koller describes the concept of physiological depth pertaining to underground plant organs as a means of understanding how they navigate through a variety of tropisms to position themselves in the soil optimally for their physiological functions. This positioning is accomplished by integrating light, temperature, gaseous, nutrient, and water signals. Deciphering this complex behavior, especially in a slow-moving organ located underground, promises to be challenging.

It is likely that every plant organ, both above and below ground, integrates a similar set of inputs. For example, when seedlings are tipped upside down, placement of a lamp below the plants does not overcome their gravitropic response; plants will reorient to grow up. The question is how? Where in the organs or organism are the effects of environmental signals collected? How are competing signals dealt with? Is there ultimately one internal signal to the motor tissues (auxin), or are there many signals that compete at the site of the motor cells? What is the hierarchy of sensing, if there is one, and how is it accomplished?

Finally, plants undergo many types of movements in response to other organisms, including trapping prey and competitive growth behaviors. Research into these activities falls in the fields of chemical ecology, plant–microbe interaction, and plant defense and plant/community ecology. A relatively well-studied example is the symbiosis of *Rhizobium* and leguminous roots. In these cases, specific bacterial and plant signals create complex movement and developmental responses in root hairs and root cortex. If the signals are recognized by receptors expressed at the surfaces of the bacteria and the legume root hair, the tip of the hair ceases its normal tip growth (similar to how a pollen tube grows) and redirects the plasma membrane to form an infection thread. The recognized bacteria enter the thread, which carries them as it grows through the root hair and into the root cortex. The thread becomes an invading organ, digesting cell walls in order to cross cells and ultimately deposit bacteria into the cortical tissue where a root nodule forms to carry out nitrogen fixation. This tightly coordinated biochemical signaling and response system, nodule formation by legume roots and *Rhizobium,* is a symbiotic result of biotic interaction. Many times, however, recognition leads to defensive responses by the plant tissues or to failure in the form of invasion and disease. Determination of a mutualistic, symbiotic, or pathogenic interaction is an active field of research, focused in large part on the mechanisms of signaling and response.

Plants live in complex communities of prokaryotic and eukaryotic species. Signaling among these organisms includes identification of beneficial versus toxic microbes, defensive signaling via volatile emission from plants, and self/nonself recognition among roots. Clouds of chemical signals create microhabitats and influence plant behaviors. In some cases, the volatile emissions of plants signal to their own leaves that danger is afoot and they should induce biochemical defenses. In other cases, the voltatiles attract insects, which in turn defend the plants from herbivores. There are also plant–plant communications manifested by movements, such as the dodder's (*Cuscuta* spp., Convolvulaceae) recognition of a suitable plant host, and its subsequent growth and twining to set up its parasitic relationship.

Movement is a macroscopic behavior, drawing our attention to a plant's ability to sense and respond to its environment. Many more responses of plants to external input are less obvious: restricted biochemical and subcellular responses, manifested perhaps in development, growth, or reproductive success (yield). Integration of the information that is inundating plant tissues and cells, coming both from the environment and consequent internal signaling, must be constantly occurring, but how? One of the big challenges facing plant biologists is to understand how hormonal, electrical, and other types of signaling elicit controlled developmental responses in plants. By having a look at plant movements and bringing out the physiological and cellular basis for them, Dov Koller has given us tools not only to imagine and explore how plants move in their environments, but, more broadly, to investigate how they live, grow, reproduce, and diversify. Given their diffuse, flexible organization, plants pose a challenging research question: How do plants integrate sensory input and coordinate biochemical, signaling, and behavioral responses? The more we consider plants to be restless, perhaps the more able we will be to form questions and test hypotheses directed toward understanding plant behaviors, including their movements.

Glossary

ABA See **Abscisic acid (ABA)**.

Abaxial Oriented away from the axis, such as the lower side of a leaf. Contrast with **Adaxial**.

Abscisic acid (ABA) Plant hormone produced under a variety of stress conditions.

Achene Single-seeded, dry fruit characteristic of the Asteraceae.

Acropetal Progression of a developmental or physiological process along an organ or tissue from base to apex. Contrast with **Basipetal**.

Actin See **Microfilament**.

Action spectrum Results from an experiment determining the relative effectiveness of monochromatic light in causing a response, according to its frequency (wavelength). Where it coincides with the *absorption spectrum* of a photoreceptor pigment, it supports the possibility that the photoreceptor is involved.

Activation energy Amount of energy invested to initiate a spontaneous process.

Active transport Mechanism for unilateral transport of ions or other solutes across impermeable cell membranes by means of transmembrane carrier proteins using metabolic energy directly (such as ATP).

Adaxial Oriented toward the axis, such as the upper side of a leaf. Contrast with **Abaxial**.

Adventitious Refers to organs equivalent to each other. Adventitious buds are shoot meristems produced on roots, and adventitious roots are those produced directly from the stem or leaf.

Aerenchyma Tissue with large parenchyma cells and intracellular air spaces.

Amyloplast A colorless plastid in the cytoplasm of certain cells that manufactures and contains starch grain(s).

Angiosperms Plants producing seeds protected within a fruit coat; also called flowering plants.

Angle of elevation (*Ae*) Angle with the vertical axis.

Anisotropic Differential changes in the dimension of an object (cell, tissue, or organ).

Anoxia Reaction to absence or abnormally low levels of oxygen.

Anther See **Stamen.**

Anthesis Flower opening.

Anticlinal Perpendicular to the external surface of an organ.

Apoplast Aqueous environment forming extracellular continuum outside the cell membrane (includes cell wall).

Apoprotein Protein component of a photoreceptor molecule.

Assimilate Synthetic products of the metabolism, especially those resulting from photosynthetic reactions.

ATP Adenosine-5'-triphosphate, the energy currency of cells that is produced by respiration.

Autotrophic Self-supporting organic nutrition, using light energy for synthesis from inorganic external resources.

Auxin Plant hormone that regulates growth, cell division, cell expansion, root branching, and apical dominance.

Auxin carriers Transmembrane proteins that transport auxin unilaterally.

Axial Along the longitudinal axis of the organ or plant.

Axial angle Angle with respect to the axis.

Axil Location between stem and adjoining leaf.

Axillary Arising from the axil.

Axonema Flexible core of flagellum or cilium.

Azimuth angle (*Az*) Angle with any predetermined reference direction.

Basipetal Progression of a developmental or physiological process along an organ or tissue from apex to base. Contrast with **Acropetal.**

Bifid Two-parted.

Bracts Nongreen leaves that may remain thin or assume protective structure for buds, either vegetative or floral.

Bulb Modified, underground stem.

Bulk modulus of elasticity Stress/strain ratio (ratio of change in pressure to fractional change in volume) in a three-dimensional object.

Bulliform cells Swollen epidermal cells oriented longitudinally along leaves of monocotyledonous plants, specialized for swelling/shrinking to cause leaf to roll/unroll during drought or other stresses.

C-terminus and **N-terminus** Locations at opposite ends of a polypeptide chain. The last amino acid residue of the C-terminus carries an a-carboxyl group, and that of the N-terminus carries an a-amino group.

Calmodulin A protein that interacts with calcium (Ca^{2+}) as a "second messenger."

Calyptra Tissue capping an organ.

Calyptrogen Meristem producing the calyptras.

Calyx Ring of sepals at the base of the flower (may be green, scale-like, white, or pigmented).

Capillarity The behavior of fluids in capillaries, determined by surface tension of the fluid.

Capillary Tube with a very small diameter; applicable also to narrow, irregularly shaped spaces.

Capitulum Inflorescence typical of the Asteraceae consisting of a common flattened disc on which numerous sessile florets are closely packed.

Capsule Dry, dehiscent fruit derived from fruit composed of one or more carpels.

Carboxylation Incorporation of carbon dioxide into a molecule. Decarboxylation is the reverse process.

Carnivorous Meat-eating.

Carpel Flower organ containing ovary or ovaries.

Cauline leaves Leaves situated on the elongated part of the shoot, above the basal rosette.

Centrifugation Rotation of mass around an axis, creating an outward centrifugal force that increases with linear velocity (revolutions per minute). Centrifugation is used to accelerate sedimentation and to simulate gravity vectors of desired magnitude.

Chemotropic response A response in a direction determined by a concentration gradient of a chemical substance.

Chimera A tissue or a substance combining two totally dissimilar elements (tissues or compounds).

Chitin Polysaccharide matrix making up cell walls in fungi, some algae, and the exoskeleton of insects.

Chromophore Molecule that absorbs light. Light absorption by the chromophore causes conformational change and activation of the attached photoreceptor protein.

Cilia Short filamentous projections from the surface of an organism, often used to move it or fluid along it.

Circadian Occurring approximately every 24 hours.

Circumnutation Describes the autonomous, advancing, spiral movements of shoot apices.

Clinostat An instrument with a platform that can be rotated at arbitrary angular velocities and at arbitrary angles with the gravity vector. With the axis horizontal, a plant attached to the platform and rotated at an appropriate angular velocity has insufficient time to respond gravitropically.

Clone Genetically identical offshoot.

Coleoptile The first embryonic leaf of the grass seedling. It differs from the succeeding foliage leaf by the absence of chloroplasts and a leaf blade (lamina); it grows into an elongated, oval sheath in the form of a scabbard, and serves as the passage for the emerging foliage leaf through the soil.

Collenchyma Tissue composed of cells with walls that are thickened at specific locations (for instance, where the wall is in contact with intercellular space).

Collimated Parallel, as in rays of light.

Columella In roots, a special group of cells that forms the core of the root cap. In certain dry fruits, the placentas of the carpels from which they are composed are joined together in the center; where the walls (septa) separating the neighboring compartments (locules) disintegrate, all that remains is a central columella on which ovules and later seeds are carried.

Compound leaf Lamina composed of a number of individual leaflets. In a pinnate leaf, they are arranged either as parallel rows along a common rachis or as branches of the rachis (rachilla), with or without a terminal leaflet. In a palmate leaf, they radiate palmately from a common juncture terminating the petiole.

Condensation Formation of a bond between two molecules in a reaction by which a hydroxyl residue of one partner and a hydrogen atom from another combine to form one molecule of water; or transformation of independent water molecules of vapor into fluid water by formation of hydrogen bonds between them as a result of their loss in kinetic energy through cooling.

Conductance The inverse of resistance to flux, such as of water (volume cm^{-2} s^{-1}) in or out of the cell.

Conformation Spatial (three-dimensional) structure and organization.

Coplanar Level—all parts on the same plane.

Corm Rounded, underground stem.

Corolla Pigmented foliar organs of the flower *(petals)* between the calyx and the stamens.

Cortex In roots, all tissues between the epidermis and endodermis. Its subepidermal layers may develop into exodermis, composed of dead cork cells with suberized walls (containing impermeable suberin, a wax-like substance).

Cotyledon First (seed) leaf of an embryo; food-storage organ.

Covalent bonding Occurs when two atoms share an electron in a common orbital.

Crassulacean acid metabolism A photosynthetic mechanism for concentrating carbon dioxide at the active site of rubisco. CO_2 is taken up at night by activity of phospho enol pyruvate carboxylase and stored as malate in vacuoles. In the day, CO_2 is released from the malate and provided to rubisco for normal photosynthetic reactions. This process occurs in crassulaceae and other families of flowering plants.

Crista(e) The internal membrane of the mitochondrion, forming numerous projections inward and thus increasing the area available for respiratory reactions.

Crozier A leaf (common among ferns) whose distal end forms a tight, flat spiral (reminiscent of a bishop's ceremonial staff) that unwinds acropetally (toward the apex) as the leaf grows.

Cryptochrome A pigment found in plants and animals that absorbs blue light. Before they were identified, this generic name was used for all photomorphogenic pigments that absorbed blue light; the name was derived from the cryptogramic organisms (such as fungi and slime molds), which were used extensively for studies of morphogenic responses to blue and ultraviolet-A light.

Cyclosis Cytoplasmic streaming.

Cytokinesis Cell division.

Cytoplasm The working contents of plant cells, internal to the plasma membrane and external to the vacuole membrane, containing the nucleus, endoplasmic reticulum, chloroplasts, and other organelles.

Cytoskeleton Network of protein fibers, including microtubules and microfilaments embedded in the cytoplasm of cells.

Cytosol The internal, fluid region of the cytoplasm.

Diaspore Dispersal unit containing seed, which also may include fruit coat, floral organs, and bracts.

Dichroic Absorbing polarized light preferentially along one plane.

Dicotyledonous Belonging to the classification of angiosperms including plants formed from seeds with two cotyledons (embryonic leaves).

Dictyosome A stack of flattened, parallel discoid sacs. Vesicles that are detached from the endoplasmic reticulum, containing its products, are added to the sack at the bottom, where they start their transformation to the finished product (primarily cell-wall components and pectin). In the course of transformation, they move up the stack. At the top of the stack, they are packed into vesicles (Golgi vesicles) that are transported through the cytoplasm to their destination, where they unload their content.

Diffusates Substances leaking out of the tissue by diffusion.

Distal Farthest. Contrast with **Proximal.**

Diurnal Cycle length equal to 24 hours.

Dorsal Side of an organ farther from the main axis from which it originates; e.g., the lower (abaxial) side of leaves often exhibiting different morphology, anatomy, or physiological responses than the opposite side. See also **Ventral.**

Dorsiventral Referring to the axis from upper to lower sides of an organ.

Elastic modulus Stress/strain ratio.

Elater Hygroscopic cell that changes shape to disperse spores in seedless plants.

Electrogenic Capable of generating electric potential.

End of day (EOD) Evening exposure to far red light transforms phytochrome to its biologically inactive form P_r, which modifies morphogenetic responses.

Endocarp See **Pericarp.**

Endodermis Innermost tissue of cortex, one cell in thickness, bordering the vascular cylinder. In the shoot, it is in the form of a starch sheath that contains amyloplasts that act as statocytes.

Endogenous Of internal origin, from within the organism itself.

Endoplasmic reticulum Membrane bound, irregularly shaped, interconnected system of flattened sacs and tubes that branches into various regions of the cytoplasm. It is the site of intensive metabolic and productive activity.

Endosperm Specialized food-storage tissue in seeds of monocots and in many other plants.

Epicotyl Section of seedling shoot above cotyledons. See also **Hypocotyl.**

Epigeal Refers to a seed whose cotyledons expand out of the seed coat and become photosynthetic. See also **Hypogeal.**

Epinastic Refers to downward curvature of an organ as a result of more growth on upper (adaxial) side. The adaxial side becomes more convex in epinastic movement; the reverse takes place on the abaxial side. See also **Hyponastic.**

Epiphyte Plant using any aerial support (dead or alive) to exist entirely without contact with the soil. Epiphytes are common in rainforests, where they use the height of their hosts to escape the deep shade.

Etiolation Developmental process leading to etiolated form. In the absence of light, shoot organs have no green (functional) chloroplasts, leaf expansion and radial growth of the stem are inhibited, and internode elongation is excessive: the pale, spindly shoot is etiolated. Exposure of etiolated shoots to light reverses these processes, causing de-etiolation.

Eukaryotic Refers to cells (and organisms containing cells) that contain organelles; nucleated.

Exine External coat of a pollen grain. The exine withstands digestion by virtually any agent, which enables it to survive as the exoskeleton of the pollen grain for millennia within all environments.

Exocarp See **Pericarp.**

Explant Part of a plant excised.

Extensibility Measure of the change in length or surface area caused by a given mechanical stress.

Extensor See **Pulvinus.**

Filament See **Stamen.**

First positive phototropic response Fluence-dependent response of the oat coleoptile to unilateral blue light, which exhibits one peak (first positive) in the low range and another (second positive) in the higher range, with different characteristics.

Flagellum Long projection from a cell that moves back and forth, propelling the cell through an aqueous medium.

Flexor See **Pulvinus.**

Fluence Total photons intercepted by unit area.

Funiculus Tissue connecting placenta to an ovule, containing conducting tissue for supply of water and assimilates.

Gametophyte Sexual (haploid) generation, producing sexual cells (gametes) in special containers (gametangia).

Geophyte Plant that lives underground.

Glume See **Spike.**

Glycolysis Decomposition of carbohydrate in anaerobic respiration.

Golgi apparatus Stacks of flat vesicles formed by the endoplasmic reticulum that manufacture most of the cellular products for export and export them toward the cell membrane in vesicles (exocytosis).

Grana stacks Stacks of grana—flattened vesicles, branches of a system of membrane-bound thylakoids inside the chloroplast. Light-harvesting complexes of chlorophyll and other photosynthetic pigments (as well as other components of the photosynthetic apparatus) are embedded in the membranes of the grana.

Gravitropic Refers to movement guided by gravity.

Gravitropic setpoint angle (GSA) Angle of elevation (Ae) of a root or shoot.

Gymnosperm Plants producing naked seeds, often referred to as conifers.

Gynophore Internode subtending the ovary.

Haustorium A specialized organ, frequently amorphous, produced by a parasitic plant, fungus, or bacterium when it invades the tissues of a plant.

Heat balance Integrated effects of elevated tissue temperature via absorption of solar radiation and sensible heat transfer from the surrounding air. Heat is lost by thermal radiation and by evaporative cooling.

Heliotropic Refers to movements guided by solar angle.

Heterosporous Producing male and female spores that differ greatly in size. See also **Homosporous.**

High-irradiance response (HIR) Response to continuous irradiation at high irradiance with far-red light ($\sim$720 nm), or to simultaneous excitation of phytochrome P_r and P_f by far-red light$_r$ that results in cycling between them and accumulation of the intermediate with the slowest rate of reaction. It is strongly dependent on irradiance.

Homosporous Producing spores that are indistinguishable externally. See also **Heterosporous.**

Hydraulic conductivity The inverse of resistance to the flux of water (volume cm^{-2} s^{-1}) in or out of the cell.

Hydrophilic Water attractant. Substances with residues (such as carboxyl, $-COOH$) attract water molecules and form hydrogen bonds with them.

Hydrophobic Water repellent. Nonpolar residues (such as $-CH_3$) are incapable of forming hydrogen bonds, specifically with water.

Hydrophyte Plant adapted to an aqueous environment, partially or fully submerged; species preferring aquatic habitats.

Hydrotropism Movement guided by water potential gradient.

Hygroscopic Ability of material to absorb or adsorb water.

Hyperpolarization Increase in the electric charge difference on opposite sides of a membrane, by convention inner-surface negative, outer-surface neutral. In plants, plasma membranes are hyperpolarized by active pumping of hydrogen (H^+) out of the cytoplasm.

Hypocotyl Section of seedling shoot below cotyledons. See also **Epicotyl.**

Hypogeal Refers to a seed whose cotyledons remain in the seed coat underground. See also **Epigeal.**

Hyponastic Refers to upward curvature of an organ as a result of more growth on the lower (abaxial) side. The abaxial side becomes more convex in hyponastic movement; the reverse takes place on the adaxial side. See also **Epinastic.**

Imbibition Absorption of water, especially in seeds before germination.

Inflorescence Cluster of flowers.

Insectivorous Referring to plants that enrich their diet with amino acids produced by digestion of the proteins in insects that they capture by a variety of mechanisms.

Integument Protective external layer of ovule.

Intercalary In between. Intercalary meristems are those that are located between or within mature tissues; intercalary growth takes place within otherwise mature tissue.

Internode Part of the stem between neighboring nodes. See also **Node.**

Irradiance Equivalent to photon fluence rate (moles photons/unit area/unit time).

Karyokinesis Nuclear division.

Kinase An enzyme that phosphorylates its substrate (attaching an energy-rich phosphate).

Lag phase Interval between start of excitation and start of response.

Lamina Leaf blade; usually flat and thin but may be cylindrical, a sector of a cylinder, an elongated pyramid, or other shapes.

Leguminous Characteristic of plants in the Fabaceae family.

Lemma See **Spike.**

Lenticel Lens-shaped pore on surface of plant that provides a path for gas exchange.

Ligand Specific atom, radicle, or molecule that attaches to the characteristic atom of a polyatomic complex.

Lumen Space, such as within cell walls or between thylakoid membranes.

Malvaceous Characteristic of plants in the Malvaceae family.

Mangrove A tree or shrub adapted to brackish or saline coastal habitats.

Mechanical advantage Antagonism ratio T_s/T_g between turgor pressure in the subsidiary cell and its neighboring stomatal guard cell. The latter has to exceed the former before stomata open.

MegaPascal (MPa) Unit of pressure ($=10$ bar).

Megasporangium In seed plants, a spore-producing structure that produces megaspores, which can become the female gametophyte (egg) and give rise, upon fertilization, to the embryo in an ovule.

Meiosis Division of a diploid cell, producing four haploid daughter cells.

Mericarp Dry, one-seeded fruit originating in an ovary with several separate carpels.

Meristem Tissue devoted to production of newly divided cells, from which mature tissues differentiate. Tissues of the root and shoot are produced by their apical meristems. Secondary conducting tissues are produced by a peripheral meristem—cambium and cork cambium (phellogen). Some meristems may remain inactive until required.

Meristematic activity Continuous cell division.

Mesocarp See **Pericarp**.

Mesocotyl First internode subtending the coleoptile in monocotyledonous seedlings.

Mesophyll Entire subepidermal tissue of leaf with the exception of the veins. It is mostly photosynthetic and characterized by an extensive system of intercellular spaces.

Microfibril Crystallized cellulose molecules that provide tensile strength to cell walls.

Microfilament Long fiber made of actin involved in scaffolding of proteins and organelles within cells.

Micromole (μmol) $=$ mole $\times 10^{-6}$.

Micropyle Access of the pollen tube. The ovule of seed plants is protected by two concentric coats that surround the nucellus, leaving the micropyle as an open pore.

Microsporangia In seed plants, a spore-producing structure that produces microspores, which can become the male gametophyte (pollen).

Microtubule Long tubular fiber made of tubulin; associated with movement of organelles within cells.

Midrib Major vein running along or close to the plane of symmetry of the lamina.

Millimole (mmol) $=$ mole $\times 10^{-3}$.

Mitochondria Subcellular organelles devoted to oxidative respiration.

Monochromatic Describes light emitted at a unique frequency, a single line in the spectrum.

Monocotyledonous Belonging to the classification of angiosperms that includes plants formed from seeds with one cotyledon.

Multicellular Composed of many cells.

Mycorrhyzae Mutalistic association of fungi with roots.

Myosin Protein that can polymerize to make filaments used for movement within cells.

N-terminus and **C-terminus** Locations at opposite ends of a polypeptide chain. The last amino acid residue of the C–terminus carries an α–carboxyl group, and that of the N-terminus carries an a-amino group.

Nanomole (nmol) $= \text{mole} \times 10^{-9}$.

Nastic Refers to movements that take place in endogenously defined spatial directions.

Node Location of stem where leaves (one or more) are attached. See also **Internode.**

Nucellus Megasporangium of seed plants, enclosed and protected by two concentric integuments.

Nutation See **Circumnutation.**

Nyctinastic Nastic movements occurring at the start of darkness.

Ontogeny Inherent course of development.

Orthogravitropic Along the gravity vector.

Orthotropic Vertical, positively gravitropic.

Osmotic Refers to behavior of solutes and water during the process of osmosis—diffusion of water across a semipermeable membrane which allows water to pass but restricts solute passage in a selective manner.

Ovary Flower part made of carpels, containing ovules.

Palea See **Spike.**

Palisade mesophyll Photosynthetic cell layer directly underneath adaxial (upper) epidermis in leaves, made up of elongated cells separated by air spaces.

Palmate Shape of leaf lamina that is divided into leaflets that radiate from a common juncture with their petiole.

Pappus Crown of hairs, bristles, or bracts on top of an achene.

PAR See **Photosynthetically active radiation (PAR).**

Parenchyma Tissue composed of relatively large, more or less isodiametric cells with a large central vacuole and little or no differentiation. The walls are commonly thin and composed primarily of cellulose, but may be suberized (impregnated with suberin), as in cork, or lignified (impregnated with lignin), as in the xylem.

Pectin Amorphous, strongly adhesive substance formed from pectic acid bound by the divalent cations calcium and magnesium. Pectic acid is a long-chain polymer of uronic acids (organic acids formed by oxidation of hexose sugars). The repeated pattern of carboxyl residues (–COOH) along the chain make pectic acid hydrophilic. In the presence of bivalent cations (Ca^{2+} and Mg^{2+}), the carboxyl residues of neighboring chains link together to form a highly adhesive pectin, the major component of the middle lamella by which neighboring cells are bonded to form tissues.

Pedicel The stalk of a single flower.

Peduncle Main axis of an inflorescence.

Perennial Living more than one year; also termed perennate.

Perianth Asexual foliar organs of the angiosperm flower—the calyx, composed of sepals, and corolla, composed of petals. When the two are indistinguishable, they are termed tepals.

Pericarp Fleshy outer layers of a fruit, including the exocarp, mesocarp, and endocarp.

Periclinal Plane of cell division parallel to surface.

Pericycle External tissue of the vascular cylinder of the root, internal to the endodermis. It remains meristematic and develops primordia of lateral roots at equally spaced azimuth angles and vertical distances.

Petal See **Corolla.**

Petiole Stem-like base of a leaf.

Phloem Conducting system for organic assimilates along pathways from the source of synthesis to the sinks where they are utilized.

Phosphorylation Addition of an energy-rich phosphate bond by condensation (removal of H^+ and OH^-). Dephosphorylation hydrolyzes a high-energy phosphate bond.

Photoequilibrium Ratio of P_{fr} (active phytochrome) formed, to total phytochrome P_{fr}/P. Because of overlap in absorption of P_r and P_{fr}, the highest achievable ratio under red light is 0.8, and the lowest achievable ratio under far-red light is <0.03.

Photoinhibition Inhibition of photosynthesis by excessive absorption of PAR, resulting in excessive photorespiration.

Photomorphogenesis Developmental processes controlled by light.

Photon Quantum of electromagnetic radiation energy.

Photonastic Refers to light-driven movements that take place in a direction determined endogenously.

Photoperiod Duration of light period during a 24-hour cycle.

Photoreceptor Molecule consisting of a chromophore, which absorbs light at specific wavelengths in the PAR/visible range (380–730 nm), bound to an apoprotein, which transduces the light energy to a chemical form.

Photosynthetically active radiation (PAR) Wavelengths of light (400–700 nm) that drive photosynthetic reactions; generally coinciding with visible light.

Phototaxis Light-driven movement of cellular organelles or free-swimming cells.

Phototropic Refers to light-guided movements that take place in a direction determined by the light vector.

Phototropin Blue-light absorbing photoreceptor.

Phragmoplast Developmental stage of the cell plate, whose growth in diameter across the equatorial plane of the cell eventually divides its protoplast during cell division (cytokinesis). At this stage, the cell plate is ringed by microtubules oriented perpendicular to it, which guide Golgi vesicles to its edges to fuse with it there. The latter are loaded with pectinaceous components of the cell plate.

Phyllotaxis Sequential acropetal (base to apex) azimuth direction of leaves on the stem.

Physiological depth Position in the soil where physiological functions operate optimally.

Phytochrome Red and far-red light-absorbing pigment. Synthesized as the inactive form, P_r, and converted to the active form, P_{fr}, by absorption of red wavelengths (660 nm). The active form P_{fr} absorbs preferentially longer wavelengths (730 nm), which cycle the molecule back to the P_{fr} form.

Pinnate Refers to a compound leaf consisting of several leaflets arranged in parallel along a common axis (rachis). A bipinnate leaf has a number of pinnae, each along its own rachilla along their common rachis.

Pit field Locations in the cell wall across which bundles of plasmodesmata connect protoplast of neighboring cells, preventing the deposition of cellulose microfibrils during formation of primary and secondary walls of both cells, thus forming opposite microscopic pits.

Placenta Tissue of carpel on which ovules develop, usually on margins of the carpel.

Plagiogravitropic Refers to plagiotropic movement coupled to the gravity vector.

Plagiotropic Growing at an acute angle to the axis of origin.

Plasmalemma Plasma membrane of plant cells bounding the protoplast (cytoplasm) and in contact with the cell wall.

Plasmodesma(ta) Cytoplasmic membrane-bound strands that pass through specialized perforations (pits) in the adjoining walls of neighboring cells, thereby transforming the individual cells of the tissue into a symplast.

Plasmolysis Contraction of the protoplast and its separation from the cell wall when the cell is in a hypertonic solution (osmotic potential more negative than cell water potential). Incipient plasmolysis is the first observable sign of plasmolysis, at which point the two are equal and the osmotic potential of the external solution is a measure of the cell water potential (of the contracted protoplast).

Plumule Apical bud and its cluster of immature leaves.

Pneumatophores Short vertical adventitious roots produced in great number by the shallow root system of mangrove plants. They protrude above the marshy surface of the saline, airless soil and provide an efficient system for passage of atmospheric oxygen through their vast aerating tissue—aerenchyma.

Pollen sac Microsporangium of seed plants.

Pollen tube Projection from the pollen grain that carries the two sperm nuclei and a tube nucleus through the flower style to the embryo sac.

Pollination Successful transfer of pollen to the stigma, as a first step to fertilization.

Polysomes Aggregations of ribosomes; sites of protein synthesis.

Presentation time Time interval between start of excitation and start of response.

Pressure probe Instrument for measuring turgor pressure in plant cells. A very thin capillary is inserted (under the microscope) into the vacuole. Turgor pressure drives the vacuolar sap into the capillary. Hydraulic pressure applied to the free end of the capillary to drive the meniscus back into the vacuole is the measure of the turgor pressure.

Primordium Plant organ during its earliest (meristematic) development, while it grows primarily by cell division.

Prop roots Thick, strong adventitious roots, produced from aerial parts of the trunk of tropical trees, that grow diagonally down into the soil, thereby supporting the tree.

Propagules Plant parts by which vegetative propagation is accomplished, such as bulbils and cormels.

Prothallium Vegetative part of the gametophyte.

Protonema Initial, filamentous stage of gametophyte (sexual generation) in mosses, liverworts, and ferns.

Proximal Closest. Contrast with **Distal.**

Pulvinus (pulvinule) Specialized turgor-mediated motor organ that connects the moving organ to its rigid support. This structure causes movement by contraction of the flexor tissue (adaxial, upper surface) to lift the leaf, and contraction of the extensor tissue (abaxial, lower surface) to lower the leaf.

Quadrifid Four-parted.

Rachis, rachilla In a compound leaf, whose lamina is composed of a number of individual leaflets situated along a common axis (rachis), or its branches (rachillae).

Radicle First root produced by germinating seedling.

Receptacle Top of flower stalk, bearing all floral organs in whorls or spirals. The shape varies: flat, cupped, sunken, raised, domed, globose, or spire-shaped. Specific internodes may elongate: below stamens or below ovary. The inferior ovary is entirely enclosed by the receptacle.

Reciprocity Describes a phenomenon where the result of a photobiological process is a function of the product of intensity (fluence rate) and duration of irradiation. It indicates that the result depends strictly on the number of molecules excited by photons and is limited within a range of intensities.

Rhizome Horizontal, underground stem.

Root cap An organ acting as a protective cap over the meristem of the root tip, formed by calyptrogen (a meristem directly opposite to the root apical meristem).

Root hairs Thin tubular outgrowths of single cells that project radially from the epidermal surface.

Rosette Describes growth habit of plants that do not elongate internodes between leaves, thus producing many leaves close to the ground.

Scale leaves Nongreen leaves that may remain thin or assume a thick, highly resistant protective structure for the buds in their axil.

Second positive phototropic response See **First positive phototropic response.**

Selective permeability Indicates resistance by a membrane to movement of molecules/ions through it by diffusion (free random movement); depends on the nature of the molecule/ion and presence of transporters in the membrane.

Sepal See **Calyx.**

Skotonastic Refers to movement driven by the light–dark transition, occurring in a direction determined endogenously.

Skototropism Movement toward darkness.

Spike An inflorescence of certain grasses in which the individual spikelets are sessile, tightly packed along a long axis. The grass inflorescence is composed of spikelets, each enclosed by one to two glumes (bracts) and containing one to several florets, each enclosed by two bracts (lemma and palea).

Spongy mesophyll Photosynthetic tissue in a leaf, with loosely-packed, irregularly shaped cells and extensive intercellular spaces, usually filling the lower (abaxial) part of the mesophyll (leaf tissues inside epidermis, excluding vascular).

Sporangium Structure containing spores.

Sporogenous Spore producing (potentially).

Sporophyll Metamorphosed leaf bearing sporangia with male spores (pollen grains) or female spores (in ovules).

Stamen The microsporophyll of the angiosperms, which characteristically consists of an elongated, slender, elastic filament terminated by an anther, which consists of two paired pollen sacs (microsporangia).

Starch sheath A single layer of cells with prominent amyloplasts separating the external cortex from the internal vascular cylinder; corresponds to the endodermis.

Statocyte Type of cell specialized for detecting the gravity vector; often contains statoliths.

Statoliths Dense particles of starch that press against the bottom of a cell.

Stigma Receptive surface of the style of a flower, where compatible pollen germinate.

Stipules Pair of appendages at the leaf base in the form of leaflets, scales, and glands.

Stoichiometry Molecular binding on a numerical basis. For instance, one FMN molecule is bound to one of the LOV domains in the blue-light photoreceptor phototropin (1:1).

Stolon Stems that run horizontally along the soil surface.

Stomate Pore created by two guard cells on the epidermal surface of plants, through which gas exchange occurs.

Strain Structural deformation resulting from application of a stress (force).

Stroma The transparent content of the chloroplast within its envelope, in which the photosynthetic thylakoids are embedded; the site of the "dark reactions" of photosynthesis.

Style Tissue connecting the stigma and ovary of a flower.

Succulent Refers to plants with thick, water-filled tissue; often adapted to dry habitats.

Symplast The integrated protoplasts of neighboring cells (two or more) that act as a unit by forming a continuum by means of plasmodesmata through a pore or pores in their walls and share a common cell membrane.

Synergid cells Two cells adjacent to the egg cell in the embryo sac of a flower.

Tap root A persistent primary root that develops from the seedling radicle, continues growing in thickness and length, and produces lateral roots that are consistently and markedly thinner.

Tendrils Formed in the position of leaves, leaflets, or branches and metamorphosed into thread-like organs of the shoot that perform spiral nutation movement, thus helping to locate support and attach to it.

Tetrapyrrole A linear molecule in which four five-member heterocyclic (N-containing) rings are attached in series by –CH=bonds.

Thermonastic Refers to endogenously controlled movement driven by temperature differential.

Thermoperiodicity A physiological response to diurnal fluctuations between a high and a low temperature.

Thigmonastic Refers to endogenously controlled movement in response to touch.

Thigmotropic Movement in response to touch in the direction dictated by the direction of the perturbation.

Thylakoid See **Grana stacks.**

Tonic effect Reinforcing influence.

Tonoplast Membrane-bounding plant cell vacuole, separating vacuole from cytoplasm.

Torque The mechanical turning moment; the product of the force and the distance from the axis on which it acts.

Transduction chain A series of biochemical and biophysical changes, each activated by its predecessor which in its turn activates its follower (often involving amplification), resulting in stepwise transformation of a perceived physical or chemical signal to a physiological response.

Transition dipole moment Determined by the location within the molecule of bonds with electronic configurations that are the most susceptible to excitation by light of the appropriate frequency (wavelength).

Tropism Movement in a direction determined by a vectorial environmental signal.

Tube cell A cell within the pollen tube with a very thin primary wall and an asexual nucleus that moves with the growing tip of the pollen tube and presumably contributes to its polar growth.

Tuber Rhizome (underground, horizontal stem) modified for storage of carbohydrate.

Turgor Positive pressure created by osmotic uptake of water into plant cells stressing the rigid cell walls.

Umbel Inflorescence typical of the Umbelliferae, in which numerous flowers originate from the same junction (pedicel). The height of the peduncles of the individual flowers may be equal, providing a globular umbel. Alternatively, the height may increase progressively toward the periphery, providing a flattened umbel.

Unicell Free-living, single-celled organism, some of which form colonies.

Vacuole A large, membrane-bound vesicle in the cell cytoplasm, filled with an aqueous solution of ions and other osmotically active substances for which its membrane (the tonoplast) regulates transport in and out. This important structure provides storage for nutrients (in some cases) or toxins (in other cases) and creates "filler" that allows plant cells to spread cytoplasm over a large surface area.

Vascular tissue Conducting tissue containing tubular vessels for long-distance transport of water, solutes, and assimilates.

Velum Thin layer of tissue covering sporangial opening.

Ventral Side of an organ closer to main axis from which it originates; e.g., the upper (adaxial) side of leaves often exhibiting different morphology, anatomy, or physiological responses than the opposite side. See also **Dorsal.**

Water free space (WFS) Space occupied by the aqueous solution outside the protoplast (walls and adjacent).

Water potential Chemical potential (free energy per mole) of water per the partial molar volume of water; water potential is expressed as pressure, and represents the net potential for water movement by diffusion.

Xylem Vascular tissue conducting water and dissolved nutrients from roots to shoots; tissue consisting largely of thick-walled, dead cells that make up wood.

Yield threshold A measure of the mechanical stress required to change the length or surface area to the limits of its elasticity, beyond which the changes become irreversible (plastic).

Further Reading

Attenborough, D. *The Private Life of Plants*. Princeton, N.J.: Princeton University Press, 1995.

Christie, J. M. "Phototropin blue light receptors." *Annual Review of Plant Biology* 58 (2007): 21–45.

Darwin, C. D., and F. Darwin. *The Power of Movement in Plants*. London: William Clowes & Sons, 1880.

Gilroy, S., and P. H. Masson, eds. *Plant Tropisms*. Ames, Iowa: Blackwell Publishing Professional, 2008.

Hangarter, R. "The Hangarter Lab." www.bio.indiana.edu/~hangarterlab/ (accessed 3/30/10).

Hangarter, R. "sLowLife: A Companion Site to the Exhibition." plantsinmotion .bio.indiana.edu/usbg/ (accessed 3/30/10).

Heide-Jorgenson, H. S. *Parasitic Flowering Plants*. Leiden, the Netherlands: Brill, 2008.

Hotta, C. T., M. J. Gardner, K. E. Hubbard, S. J. Baek, N. Dalchau, D. Suhita, A. N. Dodd, and A. A. Webb. "Modulation of environmental responses of plants by circadian clocks." *Plant Cell and Environment* 30 (2007): 333–349.

Iino, M. "Toward understanding the ecological functions of tropisms: Interactions among and effects of light on tropisms." *Current Opinion in Plant Biology* 9 (2006): 89–93.

Jaffe, M. J., and A. C. Leopold. "Light activation of contractile roots of Easter lily." *Journal of the American Society for Horticultural Science* 132 (2007): 575–582.

Juniper, B. E., R. J. Robins, and D. M. Joel. *Carnivorous Plants*. London: Academic Press, 1989.

Koller, D. "Light-driven leaf movements." *Plant Cell and Environment* 13 (1990): 615–632.

Lin, C., and D. Shalitin. "Cryptochrome structure and signal transduction." *Annual Review of Plant Biology* 54 (2003): 469–496.

Moran, N. "Osmoregulation of leaf motor cells." *FEBS Letters* 581 (2007): 2337–2347.

Nagy, F., and E. Schafer. "Phytochromes control photomorphogenesis by differentially regulated, interacting signaling pathways in higher plants." *Annual Review of Plant Biology* 53 (2002): 329–356.

Nobel, P. S. *Physicochemcal and Environmental Plant Physiology.* Amsterdam: Elsevier Academic Press, 2005.

Shimazaki, K. I. "Light-regulation of stomatal movements." *Annual Review of Plant Biology* 58 (2007): 219–247.

Simon, P. *The Action Plant: Movement and Nervous Behaviour in Plants.* Oxford: Blackwell, 1992.

Someya, N., K. Niinuma, I. Yamaguchi, and H. Hamamoto. "Circumnutation in plants." *Regulation of Plant Growth and Development* 40 (2005): 83–89.

Spalding, E. P., and K. M. Folta. "Illuminating topics in plant photobiology." *Plant Cell Environment* 28 (2005): 39–53.

Suetsugu, N., and M. Wada. "Chloroplast photorelocation." *Biological Chemistry* 388 (2007): 927–935.

Taiz, L., and E. Zeiger. *Plant Physiology.* 4th ed. Sunderland, Mass.: Sinauer Associates, 2006.

Tanaka, H., P. Dhonukshe, P. B. Brewer, and J. Friml. "Spatiotemporal asymmetric auxin distribution: A means to coordinate plant development." *Cellular and Molecular Life Sciences* 63 (2006): 2738–2754.

Terashima, I., and K. Hikosaka. "Comparative ecophysiology of leaf and canopy photosynthesis." *Plant Cell and Environment* 18 (1995): 1111–1128.

Wagner, E. "Research Group Prof. E. Wagner." www.biologie.uni-freiburg.de/data/bio2/wagner/wagner.html (accessed 3/30/10).

Went, F. W., and K. V. Thimann. *Phytohormones.* New York: Macmillan, 1937.

Whippo, C. W., and R. P. Hangarter. "Bending towards enlightenment." *Plant Cell* 18 (2006): 1110–1119.

Whitelam, G. C., and K. J. Halliday. *Light and Plant Development.* Oxford: Blackwell, 2007.

Wilkins, M. *Plantwatching.* London: MacMillan, 1988.

Wolverton, C., H. Ishikawa, and M. L. Evans. "The kinetics of root gravitropism: Dual motors and sensors." *Journal of Plant Growth Regulation* 21 (2002): 102–112.

Index